Anonymous

New Floral Guide

Autumn 1899

Anonymous

New Floral Guide
Autumn 1899

ISBN/EAN: 9783337372613

Printed in Europe, USA, Canada, Australia, Japan

Cover: Foto ©berggeist007 / pixelio.de

More available books at **www.hansebooks.com**

The CONARD & JONES CO.

ROSE GROWERS
WEST GROVE, PA.

NEW FLORAL GUIDE
·:·
·AUTUMN·

1899

The Wonderful
· NEW ·
CHENILLE PLANT.
Acalypha Sanderi
—
35 cts.

NEW
Emerald
Feather
Asparagus.
15 cts.

NEW DWARF CALLA LILY LITTLE GEM 15 cts.

TheseThree Lovely Window Plants
Only 50 cts. post paid— See Description
on next page.

BULBS, ROSES, AND OTHER BEAUTIFUL
FLOWERS FOR WINTER AND SPRING BLOOM.

WE ASK SPECIAL ATTENTION TO THE THREE ELEGANT HOUSE PLANTS SHOWN ON OUR FIRST PAGE.

They are splendid ornaments for living room or conservatory, and are among the choicest things offered this season. All flower lovers want them.

The NEW CRIMSON CHENILLE PLANT

(*Acalypha Sanderi*), or COMET PLANT. Price, 50c.. postpaid.

This is indeed a wonderful novelty—nothing like it ever seen before. It makes a neat, handsome plant, 12 to 15 inches high, has bright, glossy, green leaves, and flowers like beautiful crimson, tassels 8 to 12 inches long, looking as if made of rich **Crimson Chenille or Silken Plush.** It is very easily grown, requires no special treatment and blooms nearly all the time. When winter is past, set out in the flower bed, and it will bloom the whole season. It likes good, rich soil, all the sunshine it can get and moderate heat and moisture.

A First-class Novelty, sure to be in great demand. We send strong plants, some already in bloom. Price, 35c. each; 3 for $1.00, postpaid. Larger Size Plants, 50c. each, by express only.

EXTRA SIZE PLANTS, from six inch pots, and loaded with streaming crimson tassels. very handsome, and valuable for special decoration in churches, fairs, festivals, etc. Will retain their beauty for weeks and be a constant source of wonder and admiration to all. **Price, $1.00 each; $9.00 per Doz.,** by express only.

New Emerald Feather Asparagus.

(*Asparagus Sprengerii*).

This is one of the handsomest and most valuable Evergreen Trailing Plants we have, for house and conservatory culture. It is especially desirable for window pots, vases, baskets, etc., makes beautiful sprays of lovely green feathery foliage. Very useful for bouquets, wreaths, and all kinds of floral decorations. It is a strong, vigorous plant, easily grown, requires but little care and is a charming ornament for the parlor or conservatory. **Strong Plants, 15c. each; 3 for 40c.; Doz., $1.50 postpaid.**

New Dwarf Calla Lily—Little Gem.

(*Imported Bulbs, specially Prepared for Winter Flowering*)

The True Little Gem Calla Lily is justly esteemed one of our most beautiful winter-blooming plants, but, as grown in this country, it is sometimes slow to bloom, and we have therefore imported from Europe, especially for our Colored Plate Collection, **Extra Fine Dry Bulbs,** specially prepared for winter flowering. These will start up as soon as potted and bloom, sometimes before the leaves come.

DIRECTIONS.—Use small size pots, as Callas bloom best when pots are full of roots. When done blooming dry off and set away three or four months to rest, then re-pot in fresh earth and they are ready to bloom again. They get larger and finer with age. **Price, 15c. each; 3 for 40c.; $1.50 per Doz., postpaid. SPECIAL OFFER.**—The set of three, 1 Chenille Plant, 1 Emerald Feather Asparagus, and 1 Imported Little Gem Calla, only 50c. postpaid,or 3 Collections for $1.35.

IMPORTANT NOTICE. In reply to frequent inquiries, we would say we have no connection whatever with any other company. Do not be misled because you see the name, Dingee & Conard Co., unchanged. Neither Alfred F. Conard or Antoine Wintzer have any connection with that concern, but long since left it entirely, and though that Company still retains Mr. Conard's name because they think it their interest to do so and cannot legally be prevented, he has severed all connection with them years ago. All communications intended for Alfred F. Conard or Antoine Wintzer should be addressed to **THE CONARD & JONES CO.,** West Grove, Pa., where they will receive prompt and careful attention. Please be careful to always address

THE CONARD & JONES CO., Flower Growers, West Grove, Pa.

New Large Flowering Browallia VIOLET BLUE

This is a most charming plant for the window garden, makes nice, bushy little plants, 8 to 10 inches high and blooms all the time. The flowers are large. fully 2 inches across, and lovely deep violet blue. This is a very favorite color, and it would be hard to find a plant that is easier grown or a more constant and satisfactory bloomer. When winter is over set out in the flower bed, and it will bloom profusely the whole season. We recommend it as sure to please, both for winter and summer bloom. Very sweet and handsome. **15c. each; 2 for 25c.; $1.50 per Doz.**

THE GONARD & JONES CO.,

Floral Nurseries,

Autumn, 1899 ———————— West Grove, Pa.

Dear Friends:—

To Our FRIENDS and PATRONS.

WE thank you kindly for your liberal orders, and present herewith OUR NEW AUTUMN GUIDE, filled to overflowing with the nicest and most interesting things, all at Lowest Prices, Postpaid, to your door. Fine Roses are one of our leading specialties, but our Fall Catalog is largely devoted to Bulbs, because they must be planted in the Fall, and are by far the most beautiful and satisfactory flowers for Winter and Spring bloom. All Flower Lovers want Bulbs; hardly any other flowers so easy to grow and sure to bloom. Our Special Colored Plate Collections, offered on front and back pages of cover, are marvels of cheapness and beauty, and plenty of other equally surprising offers are scattered all through the book. People like our Low Prices and liberal Dealing, and orders are coming to us from all parts of the country. If you plant any flowers, we hope you will send us a share of your orders, and, whether large or small, you can depend on getting the Very Nicest Things at Lowest Prices. We want all to know that THE CONARD & JONES CO'S FLORAL NURSERIES, WEST GROVE, PA., is the Best and Most Satisfactory Place to get the finest Roses, Bulbs and Flower Seeds. Our guarantee goes with every order ; if anything is not right, we will make it so. We have been practicing this for Twenty Years. DON'T FORGET THE NAME.

Always address,

ALFRED F. CONARD, Prest.
S. MORRIS JONES, Treas.
ANTOINE WINTZER, Vice-Prest.
ROBERT PYLE, Secy.

THE CONARD & JONES CO.,
ROSE AND FLOWER GROWERS,
WEST GROVE, PA.

INFORMATION FOR PURCHASERS. Please use the Printed Envelope and Order Sheet whenever convenient, so as to avoid all chance of mistake, and be sure to write your full name and address clear and plain.

FREE BY MAIL. Please notice that everything offered in this book, excepting only large orders and two-year-old Roses and Shrubbery, are sent free by mail on receipt of prices given to all Post Offices in the United States. Safe arrival and full satisfaction guaranteed in every case. To Canada we send Plants, Bulbs and Seeds by mail, postpaid, but Roses and Shrubbery are prohibited. We also send to Mexico, Hawaii, Bermuda, and various European contries.

PLEASE FORWARD THE MONEY WITH THE ORDER. Remittances by P. O. order, registered letter, bank draft and express are at our risk, and we will add a handsome present to offset cost of same on all orders paid in this way. We accept new postage stamps for small amounts same as cash, but prefer money whenever convenient. When coins are sent they should be carefully wrapped in paper or cloth, and care taken to seal the letter securely. We send by express when requested, or when orders are too large to go by mail. Express charges are at the expense of the purchaser, but we add as liberally as possible to help cover the express charges.

PREMIUMS. Our prices are very low, but we allow a premium of 15 cts. on the dollar for all orders of $1.00 or upwards for Roses, Plants, Bulbs and Seeds bought at the single rate—that is, persons who send $1.00 may select to the value of $1.15, $2.00, select to the value of $2.30, and so on ; but no premium can be allowed on any of the special offers, which are already as low as they can possibly be afforded.

"HOW TO GROW FLOWERS," FREE. Besides the above, we will present absolutely free, to all who desire it, and whose orders amount to 50c. or more, a six months subscription to the new magazine, "HOW TO GROW FLOWERS," in accordance with our terms on page 28. We believe this offer is entirely unequalled and it is made for this time only.

HAYES BROTHERS CO., PRINTERS, PHILADELPHIA.

Grand New Hardy Hybrid Tea Rose = = Clara Barton

New Hardy Hybrid Tea Rose, Clara Barton. C. & J. Co., 1899.

After years of careful trial and hybridizing the most beautiful varieties, we at last succeeded in obtaining this GRAND NEW CONSTANT-BLOOMING ROSE, which has attracted so much attention, and proved of such remarkable beauty and value that we requested permission of

MISS CLARA BARTON — President of the World's Red Cross Society = =

to give it her name, in remembrance of the noble work she has done in the cause of humanity all over the world. This permission was kindly granted, coupled with the suggestion that it be called simply "CLARA BARTON." This we have therefore done, and believing that it will be found one of the most beautiful and highly valued Roses now known, we take great pleasure in introducing it to all lovers of beautiful Roses as one of the choicest varieties of which we have any knowledge. The flowers are quite large, perfectly double, and wonderfully sweet. The color is delicate flesh pink. It is a tremendous bloomer, loaded with flowers the whole season, and quite hardy. Extra fine, both for bedding and house culture. **Price, strong plants, 35 and 50c. each, postpaid. Two-year size, 75c. and $1.00 each, by express.**

THE BEST ∴ ∴ ∴ ∴ ROSES 🌿
WINTER=BLOOMING

NOTE.—Fine Roses on their own roots are one of our leading specialties ; we grow them by the hundred thousand, and offer in our Spring Guide (issued in January) all the *newest and choicest varieties* in both one- and two-year sizes. But in this, **our Autumn Guide,** we offer only a short list of the *Best Roses for Winter Bloom.*

NOTICE.—These are all extra strong, well-matured Roses, in best condition to bloom quickly and give plenty of lovely buds, and bloom all winter.

MLLE. HELENA GAMBIER—One of the best new roses; very pretty and desirable, makes a neat handsome bush, bears abundantly, large very double flowers. Color, lovely canary yellow, with deep peachy-red centre, becoming lighter as the flowers open. 15 cts. each.

THE BRIDE—One of the very best pure white everblooming roses, extra large buds and flowers of exquisite form and delightful fragrance ; creamy white, sometimes faintly tinted rose. 15 cts. each.

SOUV. DE PRESIDENT CARNOT—Extra large and beautiful, lovely sea-shell pink, tinted with golden fawn on creamy white ground ; very full and sweet. 15 cts. each.

BON SILENE—Noted for the great size and beauty of its buds; bright, rich, rosy crimson; excellent for winter bloom, and house culture. 15 cts. each.

PRINCESS BONNIE—Extra large full flowers, quite double, and deliciously sweet. The color is solid rich crimson, exquisitely shaded, constant and abundant bloomer, a hardy vigorous grower, and one of the sweetest and most beautiful ever-blooming roses you can have. 15 cts.

PERLE DES JARDINS—Soft golden yellow, the best of its color for winter forcing, and fine for open ground also ; one of the finest and most beautiful yellow tea roses. 15 cts. each.

MAMAN COCHET—This is a truly grand rose, one of the finest in the whole list. It is a vigorous grower, with rich healthy foliage. The flowers are extra large, very double, full and sweet, and borne on long stems, nice for cutting ; the color is deep coral pink, delicately tinted with silver rose, makes exquisite buds, and is deliciously fragrant ; a free and constant bloomer, and wonderfully beautiful. 15 cts. each.

CLOTILDE SOUPERT—Extra fine for house culture, one of the very best; blooms in large clusters, and is covered with buds and flowers nearly all the time, rich creamy white, with pink centre; very double and sweet. 15 cts. each.

ETOILE DE LYON—Pure deep golden yellow, very large, full and sweet; makes beautiful buds and flowers, a most constant and abundant bloomer. 15 cts.

NEW TEA ROSE, MARION DINGEE—Deep brilliant crimson, one of the darkest and richest colored everblooming roses we have; beautiful cup-shaped flowers, quite full and fragrant, and borne in great profusion; fine for house culture and winter bloom. 15c.

NEW RED PET—A very pretty miniature rose, low bushy growth, constant and profuse bloomer, small, round, very double flowers; color, deep rich red, blooms all the time, fine for pot culture and winter bloom. 15 cts. each.

QUEEN'S SCARLET—Color, rich velvety scarlet, very bright and attractive, constant and profuse bloomer, strong grower, very hardy and good. 15 cts.

SAFRANO—Always greatly admired for its beautiful buds, which are rich apricot yellow, a good healthy grower, and a quick and constant bloomer. 15 cts.

MOSELLA, or YELLOW SOUPERT—Makes a neat handsome bush, loaded with flowers all the time, never out of bloom during the growing season; medium size finely formed flowers, borne in large clusters and quite fragrant; color, pretty buff or lemon yellow, fine for bedding and house culture. 15 cts.

GOLDEN GATE—This is another grand new rose of surpassing beauty ; the flowers are of beautiful form, extra large size, very double and full, and delightfully fragrant ; ground color, rich creamy white, beautifully tinged with golden yellow, and bordered with clear rose. 15 cts. each.

THE QUEEN—An elegant rose for house culture. A most constant and abundant bloomer. Extra large flowers, very double and full, and delightfully tea-scented. Color, rich creamy white, sometimes faintly tinted with lemon yellow and pale rose; one of the very best. 15 cts. each.

THE QUEEN

REMEMBER These are extra strong re-potted roses, specially prepared for winter flowering. Price 15c. each, 2 for 25c., 4 for 50c., 8 for $1.00. Set of 16 for $2.00, postpaid, or by express, purchaser paying charges, 8 for 75c., 16 for $1.50, $8.00 per hundred.

NOTE.—Good ordinary size roses, best varieties, all labeled, 4 for 25c., 9 for 50c., 20 for $1.00, postpaid.

THE AMARYLLIS, Queen of Winter-Flowering Bulbs.

MAY BE KEPT IN POTS ALL THE YEAR ROUND OR BEDDED OUT IN SUMMER.

THE
LOVELY
PINK
AMARYLLIS
OR
BELLADONNA
LILY

FINE
BULBS
ONLY 15cts. EACH
3 FOR 40cts.
POST PAID.

COPYRIGHTED BY
CONARD & JONES CO. 1898

BELLADONNA LILY.

Pink Amaryllis or Belladonna Lily—Our illustration gives a good idea of this magnificent Amaryllis, it is lovely for house culture and Winter bloom; grand large well-expanded flowers, lovely soft rose pink, and deliciously perfumed. Extra fine large blooming bulbs, 15 cts. each.

Amaryllis Equestre or Gloriosa Lily—(See colored plate on back cover.) One of the most beautiful of this lovely family, very easy to grow and invaluable as a Winter bloomer, begins to flower almost as soon as potted. Splendid large lily-shaped flowers, 4 to 5 inches across, bright flashing orange scarlet, with lovely green and white star in centre. A most beautiful and satisfactory variety for Winter bloom; requires but little attention and may be kept in pots all the year round ; 15 cts. each.

Queen Mary or Double White Amaryllis, (Ismene Galathena). This is not a true Amaryllis, but a near relative of the family, and a most rare and beautiful flower. Particularly valuable for pot culture and Winter bloom. Throws up flower stalks 12 inches high, crowned with large, double lily-like flowers of waxy texture, pure snowy whiteness, and delicious fragrance. The outer petals are curiously reflexed and the inner ones delicately fringed. A most lovely and charming flower, entirely distinct and different from all others. Strong blooming bulbs, 20 cts. each ; 3 for 50 cts.

SPECIAL OFFER, The 3 Charming Amaryllis, only 45 cts., postpaid.

GROWS IN WATER — BLOOMS IN 2 WEEKS

THE TRUE GIANT WHITE NARCISSUS

PURE SNOW WHITE DELICIOUSLY SWEET 6 LARGE BULBS

ONLY 25 CTS. Post Paid

GIANT WHITE NARCISSUS.

Will grow in water and produce great masses of snow-white, fragrant flowers two weeks after planting; splendid for house and church decoration. We recommend all our friends to try these lovely, winter-blooming Narcissus. No other bulbs we know of will produce such an abundance of pure white, sweet scented flowers with so little trouble. A dozen bulbs, set in a bowl of water and held in place by pebbles, will make a charming pot of flowers for house or church decoration, and can be had at any time desired. The bulbs will keep in any warm, dry place till wanted, and by planting at different times, you can have a succession of lovely flowers all winter. Price, Large Blooming Bulbs, 5 cts. each, 6 for 25 cts., 50 cts. per doz.

Double Roman Narcissus.

This is one of our very finest winter-flowering varieties. The flowers are perfectly double and pure snow white, with small centre petals of rich golden yellow. It blooms very quickly either in water or soil and is delightfully fragrant. Its quick bloom and wonderful profusion of flowers is truly surprising, vast quantities are grown every season for floral decoration. Strong Blooming Bulbs, 5 cts. each, 6 for 25 cts., 50 cts. per doz.

Poeticus Ornatus.
Improved Poet's Narcissus.

This is an ideal variety for winter bloom, both on account of its exquisite beauty and remarkable earliness. Its pure waxy white flowers, with lovely crimson bordered cup in the centre, are exceedingly beautiful and always greatly admired. Sure to grow and bloom very quickly either in water or soil. 4 cts. each, 3 for 10 cts., 35 cts. per doz.

SPECIAL OFFER 2 each of the Three Varieties above, Six in all, for 25 Cts.; Four each, one doz. in all, for 50 Cts., postpaid

THE TRUE Bermuda Easter Lily.

GRANDEST OF ALL FLOWERS FOR WINTER BLOOM.

Bermuda Easter Lily.

THIS is one of the grandest and most beautiful Lilies ever seen, and the best of all for Winter flowering in pots; it grows easily and is sure to bloom. The flowers are pure snow-white, very large and fragrant, and borne in splendid clusters, six or eight at a time; vast numbers of these splendid Lilies are grown for church and house decoration every year. Plant in a pot or box which is at least six inches deep and well drained; good turfy soil with a little old well-rotted manure is best; the bulb should be covered about one inch deep; firm the earth well around it, water thoroughly, and set away in a cool place two or three weeks till the roots begin to start, then bring to the light—the window of an ordinary living room is about right; they require moderate heat, with plenty of sunshine and water, and should bloom in about three months from time of planting. Our Easter Lily Bulbs are of the very best quality and warranted to give satisfaction. We offer them in three sizes.

EASTER LILIES. Extra Large Monstrous Bulbs.—These should produce twelve to fifteen flowers each Price, 35 cts. each; 4 for $1.25; $3.50 per doz., postpaid.

EASTER LILIES. First Size Standard Bulbs.—Should produce eight to twelve flowers each. Price, 20 cts. each; 3 for 50 cts.; $1.85 per doz., postpaid.

EASTER LILIES. Good Second Size Bulbs.—12 cts. each; 2 for 20 cts.; $1.00 per doz., postpaid.

LILY AURATUM.—The gold banded Lily of Japan, considered the Queen of Lilies and the most beautiful of all, immense flowers, nearly a foot in width, borne in great clusters, seemingly more than the slender stem can bear; color rich creamy white, thickly spotted with crimson and brown, each petal having a wide golden yellow band through the centre; very fragrant and sure to bloom; exceedingly beautiful. First size good blooming bulbs, 15 cts. each; 2 for 25 cts.; $1.35 per doz. Extra size bulbs, 20 cts. each; 3 for 50 cts., $1.85 per doz., postpaid.

LILY CANDIDUM.—The Annunciation or St. Joseph Lily, next after the Easter Lily; this is one of the best lilies of all for Winter flowering in the house. It grows two to three feet high, and bears grand clusters of extra large, pure snow white flowers, very sweet and rivaling the Easter Lily in beauty. We offer good large bulbs, sure to bloom quickly. Price, 12 cts. each; 3 for 35 cts.; $1.25 per doz.

LILIUM ALBUM.—Extra large flowers, pure snow white, very sweet scented. 18 cts. each; 2 for 30 cts.; $1.75 per doz.

LILIUM ROSEUM.—One of the most beautiful of all the large, flowering Japan lilies; rose and white, spotted crimson, very handsome. 15 cts. each; 2 for 25 cts.; $1.50 per doz.

LILIUM PARDALINUM. — The California Leopard Lily. An elegant and very beautiful lily from California, rich scarlet and yellow flowers, spotted with purplish brown. This is a superb lily and always gives satisfaction, does well everywhere, blooms finely in the house in pots, and is also entirely hardy in open ground. It is one of the lilies you want. Price, 15 cts. each; 2 for 25 cts.; $1.50 per doz.

SPECIAL OFFER.
One each of the 6 Magnificent Lilies above for 85 cts., postpaid, or 2 of each, 12 in all, for $1.65.

NOTE.—When done flowering in the house these lilies may be set in open ground, where they can remain, as they are entirely hardy if given a light covering of leaves or litter.

BULBS FOR WINTER FLOWERING

OF all flowers for house culture and Winter bloom, bulbs are the most beautiful, and the easiest to grow. They are absolutely sure to bloom—**Hyacinths, Tulips, Narcissus, Crocus, Scillas, Iris, Ixias, Sparixias, etc.**, besides being entirely hardy in open ground, will bloom beautifully during Winter when kept in pots in-doors, and are in fact among the finest Winter-blooming flowers we have. Bulbs for house culture and Winter bloom should be potted as early as convenient, from September to January—any soil that is suitable for other plants will grow nice bulbs. Old, well-rotted manure is the best fertilizer. Pots or boxes of any convenient size may be used. Small bulbs can be set very close together, sometimes several in a pot ; large ones need room in proportion to their size. Do not plant too deep, one inch under ground is about right for most kinds *in-doors;*—when potted, water thoroughly and set away in a cool, dark place for two or three weeks, to rest and give the roots time to start, then take to the living room or wherever they are to remain—they don't require much heat, an up-stairs room suits them nicely. Water only when they need it, *but be sure they do not get dry at the bottom.* They will soon begin to bloom, and then their lovely flowers and exquisite fragrance will surprise and delight all who see them.

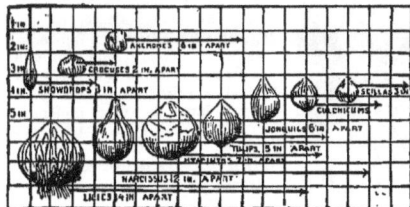

Bulb Diagram.

Bulbs in Open Ground

Tulips, Hyacinths, Narcissus, Crocus, Snow Drops, Scillas, Ixias, Sparixias, etc., are the finest bulbs for Fall planting in open ground, as they are entirely hardy and make a splendid display of gorgeous flowers very early in Spring, almost before the snow is gone. Their flowers are exquisitely beautiful and always highly valued because they come before all others. The culture is very simple, and what is better, they are absolutely sure to bloom. Other flowers may fail, but **bulbs never.**

Bulbs do not require very rich soil and will succeed well in any ordinary ground; when convenient, it is well to spade up the ground, so that it will be a little higher than the surrounding surface, and keep water from collecting on it. If the soil is poor, a liberal quantity of old, well-rotted manure should be spaded in or applied to the top as a mulch.

TIME TO PLANT. September, October, November and December are the best months for planting bulbs in the open ground. Set them from one to six inches apart, according to variety and size, and from one to three inches deep. The bulb diagram above shows the proper depth and distances apart for the different varieties.

WINTER PROTECTION. These bulbs are entirely hardy and will do without any protection, but if convenient to give the bed a light covering of leaves or litter after planting, the flowers will come earlier and be finer. The covering should be removed as soon as the plants show through in the Spring.

TREATMENT AFTER BLOOMING. When pot bulbs are done blooming they can be set away in any cool, dry place and left a few weeks to mature, after which they may be shaken out of the soil and stored away till time to plant again in the Fall. They may not make as fine flowers the second season as the first, but will usually do quite well for two or three years. Bulbs in open ground, when done blooming and well matured, may be lifted and dried off, and then treated exactly like those from pots.

Feathered or Cockade Hyacinths

Feathered or Cockade Hyacinths— Lovely and curious little flowers with feathery, plume-like spikes, deep blue tinged with red, fine for pot culture and bedding out also, hardy. 3 for 10 cts., per doz., 25 cts.

Grape Hyacinths—Pretty spikes of lovely rich, blue bell-shaped flowers ; fine for bedding and pots. 2 for 5 cts., per doz., 15 cts.

Snow White—A very scarce sort. 3 for 10 cts., per doz., 25 cts.

Special Offer.

5 Feathered Hyacinths, 5 Grape Hyacinths, and 2 Snow White, 12 in all, postpaid, for - - - - **25c.**

Feathered or Cockade Hyacinths.

THE BEST HYACINTHS.

(SEE CULTURAL DIRECTIONS, PAGE 9.)

HYACINTHS are the most useful and popular of all hardy bulbs. They come in many lovely shades, and are exceedingly beautiful and fragrant. When planted in open ground in the Fall they bloom splendidly very early in Spring, and for house culture in pots they surpass all other flowers in exquisite beauty and delightful fragrance. They are among the easiest of all flowers to grow, and are absolutely sure to bloom.

SPECIAL NOTICE.—We send good, strong, well-ripened bulbs only, such as will be sure to bloom and produce large fragrant flowers of brightest colors. The prices named include the postage, which is paid by us. You can depend on getting the best goods at lowest prices. We want your orders, whether large or small, and will see that you get liberal value and prompt attention.

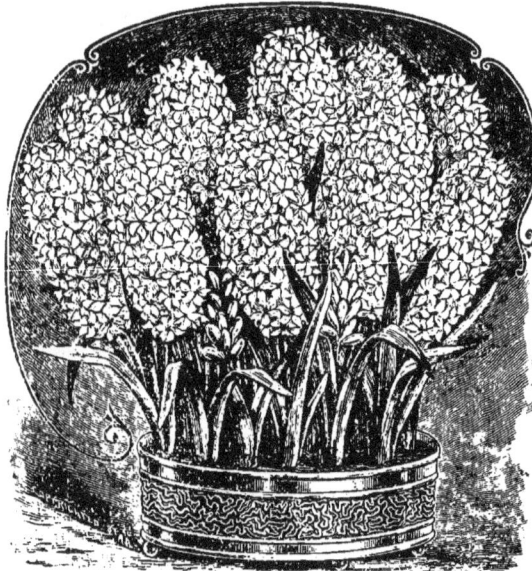

A Pan of Dutch Hyacinths for Winter Bloom.
The 12 for $1.10, Postpaid.

Single Named Hyacinths.

12c. each, $1.10 per doz., postpaid.

Amy—Fine spikes of medium size flowers, deep glowing carmine, very rich and beautiful. 12c. each.

Gertrude—Extra fine flowers in large full spikes, lovely pink passing to silver rose, striped carmine. 12c. each.

Lord Macauley—Rich bright red, with white centre, extra large spikes, very striking and beautiful. 12c. each.

Baron van Thuyll—(Blue.) Splendid large spikes of dark blue flowers, very handsome. 12c. each.

Baron van Thuyll—(White.) Pure snow white, large compact trusses. A fine early bloomer. 12c. each.

Charles Dickens—Fine spikes of extra large flowers, tender blush passing to pink, delicately shaded violet. 12c. each.

King of Blues—Beautiful large trusses of elegant dark blue flowers. Very handsome. 12c. each.

Fleur d'Or—A lovely shade of pure golden yellow, pretty graceful spikes, very attractive. 12c. each.

Mont Blanc—Pure creamy white, grand spikes of lovely large flowers, very sweet. 12c. each.

La Franchise—Rich creamy white, faintly tinted with canary yellow, large waxy flowers in heavy spikes. 12c. each.

Grandeur à Merville—Lovely creamy white, tinged with delicate blush, large trusses, exquisite kind. 12c. each.

Herman—Beautiful orange yellow, large full spikes, very handsome and sweet. 12c. each.

SPECIAL OFFER. The complete set of Twelve Best Single Hyacinths for $1.10, postpaid, or Six for -- 55c.

Fine Single Hyacinths in Separate Colors.

FIRST-CLASS BULBS AT BARGAIN PRICES.

We ask special attention to our fine mixed Hyacinths in separate colors, they are first-class selected bulbs, warranted to bloom finely and give excellent satisfaction for Winter bloom in window-garden or greenhouse, and also for bedding in open ground.

PURE WHITE.—Lovely large spikes of fine pure white flowers, very sweet. Price, 7 cts. each, 3 for 20 cts., 12 for 65 cts., postpaid. By express, $5.00 per 100.

BLUSH WHITE.—Lovely creamy white, delicately tinged with soft rosy blush, exceedingly beautiful. Price, 7 cts. each, 3 for 20 cts., 12 for 65 cts., postpaid. By express, $5.00 per 100.

DARK RED.—Rich crimson flowers, large full spikes, very handsome. Price, 7 cts. each, 3 for 20 cts., 12 for 65 cts , postpaid. By express, $5.00 per 100.

ROSE AND PINK.—These are among the prettiest of all, good bloomers and very sweet. Price, 7 cts. each, 3 for 20 cts., 12 for 65 cts., postpaid. By express, $5.00 per 100.

DARK BLUE.—Dark rich indigo blue, some almost black. Always greatly admired, and contrasts finely with other colors. Price, 7 cts. each, 3 for 20 cts., 12 for 65 cts., postpaid. By express, $5.00 per 100.

LIGHT BLUE.—Lovely violet or porcelain blue, very beautiful and desirable. Price, 7 cts. each, 3 for 20 cts., 12 for 65 cts., postpaid. By express, $5.00 per 100.

YELLOW.—This is always scarce, but very beautiful and attractive. Price, 7 cts. each, 3 for 20 cts., 12 for 65 cts., postpaid. By express, $5 00 per 100.

SINGLE HYACINTHS—ALL COLORS MIXED. A fine mixture of different colors and shades, good, sound, blooming bulbs, fine for bedding out. Price, 6 cts. each, 12 for 60 cts., postpaid. By express, $4.50 per 100.

WHEN SENT BY EXPRESS THE PURCHASER PAYS THE EXPRESS CHARGES.

Best Double 🌀 🌀 Named Hyacinths

The Prices Given are for Large Sound Bulbs, Postpaid.

BOUQUET TENDRE—Beautiful medium-sized semi-double flowers, in fine compact trusses; color, exquisite rosy blush, extra fine. 12 cts. each.

GROOTVORST—Extra large, handsome spikes of fine rosy pink flowers; very double; a superb kind. 12 cts. each.

GARRICK—A truly magnificent sort, lovely azure blue flowers, in large compact spikes; one of the very best. 12 cts. each.

CROWN PRINCE CHARLES—Dark violet blue, large full spikes of exquisite flowers; very double and sweet. 12 cts. each.

LA VIRGINITE—Fine medium-sized truss, large double flowers; rich creamy white, delicately shaded to blush. 12 cts. each.

LORD WELLINGTON—Clear bright carmine, extra large and handsome in every way; one of the very best. 12 cts. each.

BLOCKSBERG—Lovely porcelain blue, fine double flowers in large trusses; very beautiful. 12 cts. each.

LA TOUR D'AUVERGNE—Double pure white, very rich and waxy, blooms early, is one of the best kinds for all purposes. 12 cts. each.

SUPREME YELLOW—Pure rich golden yellow, very double and fine; the best double yellow. 12 cts. each.

PRINCE OF WATERLOO—Extra fine double white flowers in large full spikes; deliciously sweet. 12 cts. each.

Special Offer The complete set of 10 Best Double Hyacinths for 90c., postpaid, or 12 for $1.10.

Double Mixed 🍃 🍃 Hyacinths

Our Double Mixed Hyacinths are strong heavy Bulbs, warranted to give satisfaction.

DOUBLE DARK RED—Rich crimson flowers, large full spikes, very handsome. Price 7 cts. each, 3 for 20 cts., 12 for 65 cts., postpaid. By express, $5.00 per 100.

DOUBLE ROSE AND PINK—These are among the prettiest of all, good bloomers, and very sweet. Price 7 cts. each, 3 for 20 cts., 12 for 65 cts., postpaid. By express, $5.00 per 100.

DOUBLE DARK BLUE—Dark rich indigo blue, some almost black, always greatly admired and contrast finely with other colors. Price 7 cts. each, 3 for 20 cts., 12 for 65 cts., postpaid. By express, $5.00 per 100.

DOUBLE LIGHT BLUE—Lovely violet or porcelain blue, very beautiful and desirable. Price 7 cts. each, 3 for 20 cts., 12 for 65 cts., postpaid. By express, $5.00 per 100.

DOUBLE PURE WHITE—Lovely spikes of fine pure white flowers, very sweet. Price 7 cts. each, 3 for 20 cts., 12 for 65 cts., postpaid. By express, $5.00 per 100.

DOUBLE BLUSH WHITE—Lovely creamy white, delicately tinged with soft rosy blush, exceedingly beautiful. Price 7 cts. each, 3 for 20 cts., 12 for 65 cts., postpaid. By express, $5.00 per 100.

DOUBLE YELLOW—This variety is scarce, but always greatly admired, very sweet and lovely. Price 7 cts. each, 3 for 20 cts., 12 for 65 cts., postpaid. By express, $5.00 per 100.

DOUBLE HYACINTHS—All colors mixed, An excellent mixture of the most desirable colors and shades, fine for bedding out. Price 6 cts. each, 12 for 60 cts., postpaid. By express, $4.50 per 100.

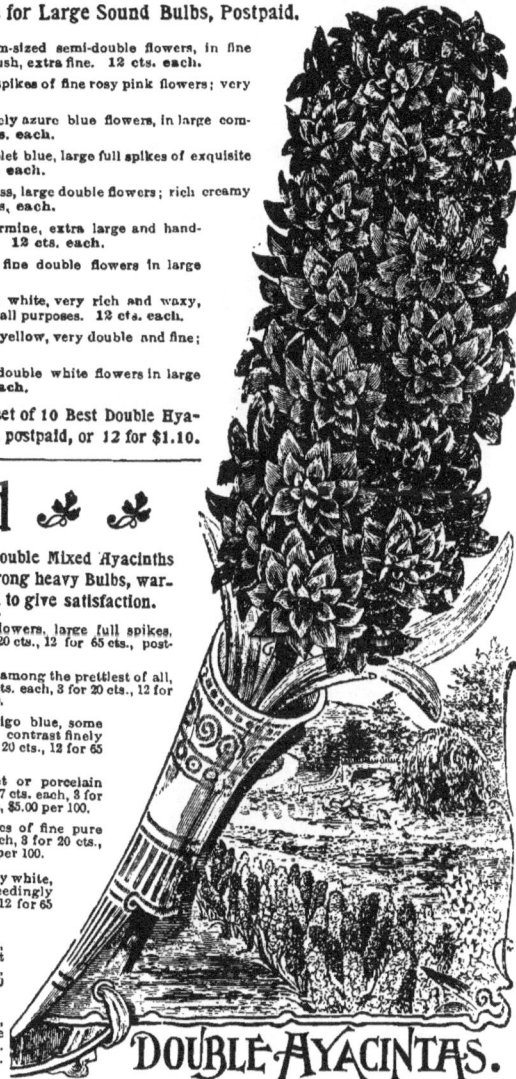

DOUBLE HYACINTHS.

Hyacinths in Open Ground should be planted 6 inches apart. A bed six feet by three feet will hold 75 Hyacinths nicely. Price, by express, purchaser paying express charges, $3.75.

A circular bed, five feet in diameter, holds 100 hyacinths. Price, by express, purchaser paying charges, $5.00.

NOTE.—Five separate colors, one ring each, make a nice combination for a circular bed. If the bed is five feet in diameter, the inside or centre ring will require 9 bulbs, the second ring 14, the third ring 20, the fourth ring 25, the fifth or outside ring 32—100 in all. Please select colors desired, or if preferred we will select for you and guarantee to please.

Collections of Hyacinths for Beds.

Circular Bed of Hyacinths.

HYACINTHS must be planted in the Fall, and are esteemed among the most beautiful of all Spring-flowering bulbs. The flower spikes are so regular in form and the colors so bright and distinct, that it is easy to make any design you wish flash forth in glowing colors at the advent of Spring. Our **Selected Hyacinths in separate colors** are largely used for this purpose, as they are sure to bloom satisfactorily, and the price is as low as can possibly be afforded.

The Circular Bed shown above is a very favorite design. It is 6 feet in diameter, and requires 108 Hyacinths, selected as follows, beginning three inches from the edge :

The **1st, or** outside row, requires **33** Hyacinths, color, deep red.

The **2nd** Row requires **27** Hyacinths, color, pure white.

The **3rd** Row requires **21** Hyacinths, color, blue.

The **4th** Row requires **15** Hyacinths, color, blush.

The **5th** (centre) Row requires **12** Hyacinths, color, purple.

DIRECTIONS.—Plant in open ground as early as convenient, from October to December ; set the bulbs so that the tops will be three or four inches under the surface and six inches apart each way, and give a light covering of leaves or litter during Winter. **The price of this Bed—108 Hyacinths in all—selected as above, is $5.40.** Packed to express here. A Circular Bed, 4½ feet in diameter, holds 54 Hyacinths. **Price, $2 70,** purchaser paying express.

Roman Hyacinths, extra fine for Winter Blooming

Roman Hyacinths
Extra Fine for Winter Flowering.

Roman Hyacinths are among the handsomest and most desirable of all bulbs for early Winter flowering. They begin to bloom very quickly and throw up a great mass of sweet and lovely flowers almost before other bulbs get started. When planted early they can easily be had in bloom by Christmas, and will continue in perfection for several weeks, if not kept too warm. They come in lovely shades of white, pink, blue and yellow. The flowers are borne gracefully on tall stems, and are delightfully sweet. They grow easily in pots, boxes or glasses, require no special treatment and are absolutely sure to bloom. Though recommended for house culture, they also do well in open ground, and will bloom very early in Spring.

SINGLE PURE WHITE.—Early and profuse bloomer, very beautiful and sweet. 5 cts. each ; 12 for 45 cts.

SINGLE BLUSH WHITE.—Rich, creamy white, elegantly tinted with rosy blush; very beautiful; immense bloomer. 5 cts. each; 12 for 45 cts.

SINGLE PINK.—A standard sort, of great beauty and delicious fragrance. 5 cts. each; 12 for 45 cts.

DOUBLE PINK.—An excellent new variety; beautiful large flowers, in handsome, well-filled spikes. 6 cts. each; 3 for 15 cts.; 12 for 50 cts.

SINGLE CANARY YELLOW—New and very beautiful; stock limited. 7 cts. each; 3 for 20 cts.; 75c. per doz.

SPECIAL OFFER.—Complete set of 5 Named Roman Hyacinths offered above, for 25 cts., postpaid.

New MINIATURE or POMPON Hyacinths.

These charming little Hyacinths are greatly admired, both for Winter bloom indoors and for bedding out ; they resemble the Dutch Hyacinths, but are smaller in size, and besides costing less, can be planted close together in pots, boxes or beds. The bulbs often produce three or four flower spikes each, and can be depended on to make a lovely display very quickly ; they are deliciously sweet and sure to please. Do not fail to give them a trial. We offer them in five separate colors,

Pure White, Rose Pink, Red and Crimson, Dark Blue and Light Blue. Price, 4 cts. each, 3 for 10 cts. The 5 colors for 15 cts.; 35 cts. per doz., postpaid.

Collections of Tulips For Beds.

CIRCULAR BED OF TULIPS.

Tulips are the most showy and handsome of all Spring-flowering bulbs. No other flowers can equal them in brilliant and gorgeous colors ; they are entirely hardy, need no protection, and are absolutely sure to bloom. Our Bedding Tulips are specially selected to grow the same height and bloom at same time. We have them in Pink, White Crimson, and Yellow. The bed illustrated above is six feet in diameter and holds 156 tulips planted five inches apart each way. The bed is divided into four sections, as indicated ; each section holds 39 tulips of same color—39 Red, 39 White, 39 Pink, 39 Yellow. Price for the 156, packed to express here, $1.25, or $1.80 prepaid through. Price per 100, 80 cts., purchaser paying charges ; or $1.15 per 100, prepaid through.

DOUBLE HERBACEOUS PAEONIAS.

These noble flowers bloom quite early in the Spring, their size is immense, are perfectly double, and borne in great profusion as regularly as the seasons come. They are of **various colors, White, Pink, and Crimson.** The roots are entirely hardy, and live on from year to year, one of the grandest hardy flowers in cultivation. As their roots are heavy and bulky, the price is higher. **Fine Mixed Colors only 35c. each ; Two for 60c., Postpaid. By Express, 25c. each ; $3.00 per Dos.**

NEW DOUBLE FLOWERING NASTURTIUMS.

Two Elegant Varieties, Novel and Beautiful.

Sunbeams.—Bright golden yellow, with carmine markings, very pretty and desirable. **15c.**

Sunset.—Color, rich orange crimson, with markings of Indian red, handsome and attractive. **15c.**

The flowers are large, often over two inches across and as double as roses ; they have a rich spicy fragrance, and the colors are so bright and striking they attract a great deal of attention. They bloom continuously and are greatly admired for pots, window boxes, baskets, vases, etc. **Price, 15c. each, the two for 25c., postpaid.**

The Best Single Early Tulips

Single Early Tulips

Best Named Varieties

TULIPS are well-known as the most showy and handsome of all Spring-flowering bulbs. Nothing can equal them in dazzling beauty. They are entirely hardy and equally valuable for planting in open ground and for house culture in pots or boxes. They do well everywhere, and whether planted in rich ground or poor, sunshine or shade, can always be depended upon to throw up their gorgeous flowers. Our list includes the very best varieties for home planting.

Set of 15 Tulips named below for 40c., 3 of each, 45 in all, $1.00, postpaid.

DUC VAN THOL—Yellow. 4 cts. each, 3 for 10 cts., per doz., 40 cts.

KAIZER KROON—Red, gold and yellow. Splendid. 3 cts. each, 3 for 8 cts., per doz., 25 cts.

POTTEBAKER—Pure Yellow. Extra large and fine. 3 cts. each, 3 for 8 cts., per doz., 30 cts.

POTTEBAKER—Scarlet Crimson. Extra large and fine. 3 cts. each, 3 for 8 cts., per doz., 30 cts.

POTTEBAKER—White. Extra large and fine. 4 cts. each, 3 for 10 cts., per doz., 35 cts.

PROSERPINE—Rich, satiny rose. Extra large and fine. 4 cts. each, 3 for 10 cts., per doz., 33 cts.

YELLOW PRINCE—Splendid rich yellow, extra large and fine. 3 cts. each, 3 for 8 cts., per doz., 35 cts.

VERMILLION—Brilliant scarlet. Extra large and fine. 4 cts. each, 3 for 10 cts., per doz., 35 cts.

LA REINE—Pure white. Extra large and fine. 3 cts. each, 4 for 10 cts., per doz., 25 cts.

Single Tulips.

BELLE ALLIANCE—Very beautiful rich scarlet. 3 cts. each, 3 for 8 cts., per doz., 30 cts.

CHRYSOLORA—Beautiful pure yellow, extra fine. 3 cts. each, 3 for 8 cts., per doz., 30 cts.

COTTAGE MAID—Splendid carmine pink, centre of petals feathered white, with yellow base. 4 cts. each, 3 for 10 cts., per doz., 35 cts.

DUC VAN THOL—Carmine. Very early and handsome. 3 cts. each, 4 for 10 cts., per doz., 25 cts.

DUC VAN THOL—Scarlet. 3 cts. each, 4 for 10 cts., per doz., 25 cts.

DUC VAN THOL—White. 3 cts. each, 3 for 8 cts., per doz., 30 cts.

C. & J. SUPERFINE SINGLE MIXED TULIPS FOR BEDDING

OUR SUPERFINE MIXED TULIPS are made up from the best named kinds of the brightest and most desirable colors for bedding, selected to bloom at the same time and grow the same height. These are the finest Bedding Tulips to be had, and are warranted to please. Price, 5 for 10 cts., 12 for 20 cts., 25 for 40 cts., 100 for $1.15, postpaid. 80 cts. per 100, purchaser paying charges.

NOTE. Tulips should be planted 4 to 5 inches apart. A circular bed 5 feet across, or 15 feet in circumference, holds 200 tulips. Price $1.15 per 100, postpaid, 80c. per 100 by express at purchaser's expense.

The Best ❧

Double ❧ ❧

Early Tulips

DOUBLE TULIPS

COPYRIGHTED 1893

Named Varieties
Double Tulips

DUKE OF YORK—Bright rose, bordered white. 3 cts. each, 3 for 8 cts., per doz., 25 cts.

GLORIA SOLIS—Deep crimson, with broad golden margin. 3 cts. each, 3 for 8 cts., per doz., 25 cts.

LA CANDEUR—Pure white, extra large and full. 3 cts. each, 3 for 8 cts., per doz., 25 cts.

MURILLO—Blush white, shaded rose. 4 cts. each, 3 for 10 cts., per doz., 40 cts.

SALVATOR ROSA—Magnificent double, pink flamed white. 5 cts. each, 3 for 15 cts., per doz., 50 cts.

SCARLET KING—Grand rich glowing crimson. 3 cts. each, 3 for 8 cts., per doz., 30 cts.

REX RUBRORUM—Bright crimson scarlet. 3 cts. each, 3 for 8 cts., per doz., 30 cts.

TOURNESOL YELLOW—Bright golden yellow, shaded orange. 5 cts. each, 3 for 15 cts., per doz., 50 cts.

TOURNESOL CRIMSON—Bordered yellow. 4 cts. each, 3 for 10 cts., per doz., 35 cts.

DUC VAN THOL—Double red and yellow. 3 cts. each, 3 for 8 cts., per doz., 25 cts.

Set of 10 Double Early Tulips, postpaid, for 30c., 3 of each, 30 in all, for 85c.

DOUBLE EARLY TULIPS—Best Varieties Mixed.

Price, 2 for 5c.; 25c. per doz.; $1.20 per 100, postpaid; 85c. per 100, Exp., buyer paying charges.

DOUBLE LATE-FLOWERING TULIPS.

These Grand Varieties bloom a little later than the single early kinds.

ADMIRAL KINGSBERGEN—Red and yellow, very handsome. 3 cts. each, 3 for 8 cts., per doz., 25 cts.

BLUE FLAG—Rich purplish blue, quite distinct. 3 cts. each, 3 for 8 cts., per doz., 25 cts.

BELLE ALLIANCE—Blue and white, extra fine. 4 cts. each, 3 for 10 cts., per doz., 30 cts.

VIOLET PICOTEE—Violet, bordered and tinted with white. 3 cts. each, 3 for 8 cts., per doz., 25 cts.

MARRIAGE DE MA FILLE—Pure white, feathered with rich crimson. 5 cts. each, 3 for 12 cts., per doz., 45 cts.

PEONY GOLD—Rich crimson and golden yellow, magnificent double. 3 cts. each, 3 for 8 cts., per doz., 25 cts.

PEONY RED—Resembles a grand blood-red Peony, extra fine. 3 cts. each, 3 for 8 cts., per doz., 25 cts.

ROSE CROWN.—Brilliant rose, very large, full and double 3 cts. each, 3 for 8 cts., per doz., 25 cts.

YELLOW ROSE—Pure yellow, the true yellow rose. 3 cts. each, 3 for 8 cts., per doz., 25 cts.

COUNT OF LEICESTER—Rich crimson and white. 3 cts. each, 3 for 8 cts., per doz., 25 cts.

Set of 10 Double Late Tulips, 25c., postpaid, or two of each, 20 in all, for 50c.

Double Late Flowering Tulips—Finest Colors Mixed, price by mail, postpaid, 2 for 5c., 12 for 25c., 100 for $1.20. By Express, at purchaser's expense, 85c. per hundred.

TULIPA GREIGI.

THE ROYAL TULIP also called THE QUEEN OF TULIPS.

THIS is a rare and costly variety, but is as easily grown as any tulip, and requires the same treatment, whether in pots or open ground. The large handsome foliage is curiously spotted with dark rich maroon, the flowers are remarkably rich and handsome ; color, brilliant orange scarlet, with yellow and black centre, undoubtedly the grandest of all tulips. 15c. each ; 2 for 25c.; $1.50 per doz.

Tulipa Greigi.

Variegated Foliage Tulips.

VARIEGATED FOLIAGE TULIPS have broad, wavy green leaves, elegantly striped and variegated with rich creamy white, giving the plant a highly ornamental appearance. Very handsome, both for beds and pot culture; they bloom early, are always bright and handsome, and the price is very low.

LAC VAN RHIJN.—Dark violet red, bordered with pure white. Very rich and handsome. 4 cts. each, 3 for 10 cts., 35 cts. per doz.

PURPLE CROWN.—Splendid large flowers, dark purplish red. Very bold and striking. 4 cts. each, 3 for 10 cts., 35 cts. per doz.

SILVER STANDARD.—A grand variety; pure white, crossed with rich crimson. Handsome foliage. 4 cts. each, 3 for 10 cts., 35 cts. per doz.

The set of 3 Variegated Foliage Tulips, only 10c., or 4 of each, 12 in all, postpaid, for only **35 cts.**

SWEET-SCENTED TULIPS FOR POT CULTURE.

We want our friends to try these lovely New Sweet-scented Tulips—they are extra fine for House and Conservatory Culture, and also for bedding out—they grow easily and never fail to bloom, and the flowers are very rich and handsome and deliciously fragrant.

GOLDEN PEARL. (Newton)—An exquisite new variety ; the sweetest of all sweet-scented tulips ; has the delicious tea fragrance of American Beauty Rose, lovely golden yellow blossoms, blooms very early; the finest sweet-scented tulip yet introduced. A gem. 10c. each, 3 for 25c., 6 for 40c., 80c. per doz., postpaid.

PRINCE OF AUSTRIA.—Grand flowers of largest size, thick glossy petals, rich orange scarlet, delightfully sweet scented. 5 cts. each, 3 for 12 cts.; 40 cts. per doz.

FLORENTINA ODORATA.—The Tea Rose Tulip—Rich yellow flowers—beautiful buds, very sweet. 5 cts. each, 3 for 12 cts.; 40 cts. per doz.

MACROSPILA.—Brilliant citron red, with dark clouded centre, finely bordered with bright golden yellow, very large and handsome. 5c. each, 3 for 12c.; doz.. 40c.

The set of 4 varieties for 20 cts.

CHOICE MAY FLOWERING Garden Tulips

BOUTON D' OR.—Deep rich golden yellow, handsome globe shaped flowers, exceedingly beautiful. 3 cts. each, 35 cts. per doz., postpaid.

BRIDESMAID.—Bright rich scarlet, striped pure white. Very distinct and beautiful. 4 cts. each, 40 cts. per doz.

GOLDEN EAGLE.—Extra large and handsome, rich orange yellow flowers, each petal edged with bright crimson, very showy. 3 cts each, 30 cts. per doz.

MAY BLOSSOM.—Pure white, elegantly striped and variegated with red, finely formed and a most charming variety. 4 cts. each, 45 cts. per doz.

ELEGANS.—Very grand and showy, rich crimson scarlet, large handsome flowers. 4c ; doz., 40c.

WHITE SWAN.—A splendid pure white tulip, extra large flowers, with broad silken white petals. Very handsome. 3 cts. each, doz.. 30 cts.

The set of Six Charming Garden Tulips, postpaid, only . . . **20 cts.**

New Picotee Tulips.

Most beautiful of all and called Picotee because the petals are elegantly feathered crimson. Extra fine for house culture and also for the garden.

PICOTEE WHITE.—Large ribbon-like petals, pure waxy white with feathered crimson border, 5 cts. each., 40 cts. per doz., postpaid.

PICOTEE YELLOW.—Deep golden yellow, exquisitely feathered with rich crimson, 5 cts. each, 40 cts. per doz., postpaid.

A FEW SELECTED GEMS.

Garden or Picotee Tulips.

Giant Gesneriana Tulips.

The Gesneriana Tulips are among the very handsomest of all, they grow 1½ to 2 feet high, and are very grand and stately; they show large, deep green leaves, and throw up tall, graceful flower stalks which bear great rich blooms, larger than tea cups. Color, bright, glowing, crimson scarlet, each petal showing a deep blue-black blotch at the base; exceedingly rich and handsome. They are extra fine for bedding, as they come after the earlier kinds are gone, and make a grand display of splendid color for several weeks. They are entirely hardy, and will bloom on from year to year. We recommend all our friends to try a bed of the True Gesneriana Tulips. Their brilliant beauty is truly marvelous.

Giant
Gesneriana
Tulip.

Price 3 cts. each, 3 for 8 cts., 30 cts. per dozen. $1.85 per 100, postpaid. By express, $1.50 per 100, buyer paying charges.

YELLOW GESNERIANA or GOLDEN CROWN— A magnificent golden yellow Gesneriana, with orange markings, makes a grand combination when planted with the Scarlet Gesneriana. Price 3 cts. each, 3 for 8 cts., 30 cts. per dozen.

Parrot or Dragon Tulips ❧

These are very curious and remarkable, they belong to the late or May-flowering Tulips. The flowers are very large, frequently six or seven inches across, with petals deeply toothed, fringed and twisted in the most striking and fantastic manner, sometimes representing the head and beak of a parrot. The colors are exceedingly brilliant and showy, crimson and yellow, flaked, dashed and feathered with green, gold and scarlet. Fine for bedding and to plant among shrubbery. They grow ten inches high and always attract attention.

Parrot Tulips.

	Each.	Per doz
Belle Jaune—Pure deep yellow,	$ .03	$ .3˚
Cafe Brun— Coffee color and yellow,	.03	.3¾
Constantinople—Deep blood red,	.03	.30
Feu Brilliante—Rich satiny crimson,	.05	.40
Mark Graaf—Crimson and orange, very handsome,	.03	.30
Orange (Gloriosa)—Orange and crimson variegated,	.04	.35
Perfecta—Red and yellow, extra fine,	.03	.30
Fine Mixed Parrot Tulips—$1.50 per 100,	.03	.25

Set of Seven Parrot Tulips postpaid, for   -   -   -   **20c.**

TWO BEAUTIFUL IRIS.

The English Iris.—Large, handsome flowers, borne on stout stems, 18 to 20 inches high; color, rich purple, violet blue and lilac; fine for beds and borders, perfectly hardy; needs no protection. Mixed colors. 3 for 10 cts; 25 cts. per dozen; $1.50 per hundred, postpaid.

The Orris Root Iris. (*Iris Florentina.*)—This lovely Iris is always greatly admired. It is a slender, graceful plant, the flowers are exquisitely formed and deliciously scented, the color is rich violet blue, feathered with bronze yellow. The delightfully scented "Orris Root" is produced from this plant. Price, 10 cts. each; $1.00 per dozen, postpaid.

NEW JAPANESE ❧ ❧ ABUTILON SAVITZII.

OUR illustration gives a good idea of the Elegant New **Japanese Abutilon, Savitzii.** It is a very striking and beautiful plant, entirely different from all others, and especially desirable for house and conservatory culture. **15 cts. each.**

New Abutilon, Infanta Eulalia.—A most charming plant for window and house. Plants grow low and compact and produce exquisite large cupped flowers in wonderful profusion. Color, lovely soft satin pink. Price, 15 cts. each.

New Abutilon, Snow Ball.—This is a beautiful pure white variety, with deep, cup-shaped flowers and handsome foliage. A true perpetual bloomer. Best white, 10 cts. each.

SPECIAL OFFER   The 3 Abutilons for **35 cts., postpaid.**

New Japanese Abutilon Savitzii.

New Carex Japonica.

A FIRST class novelty and an excellent plant for house and table decoration; forms a mass of erect-growing, fine spray-like foliage, drooping gracefully from the centre; the delicate dark green leaves, beautifully edged, as with a band of gold. **15 cts.; 2 for 25 cts.; larger size, 20 cts.**

New Carex Japonica.

Beautiful Dwarf OTAHIETE ORANGE.

THIS beautiful dwarf orange tree is one of our very prettiest house plants; thick glossy green leaves and deliciously sweet, pure white flowers; begins fruiting at once, and bears flowers and lovely golden yellow oranges all the year round; easily grown and requires very little care. **Nice plants, 15 cts. each; larger plants, 20 and 25 cts.,** according to size.

GOLDEN STAR OXALIS, (ORTGIESI.)

A charming house plant; quite rare, but always a favorite wherever seen; grows neat and compact in small tree form; dark olive green leaves, bright crimson on the under side, and bright yellow star-shaped flowers, borne in clusters all the year round. **15 cts** each; 2 for 25 cts.

Dwarf Otahiete Orange.

Narcissus or Daffodils.

OUR illustration aims to give some idea of the different varieties of the beautiful Narcissus described below. They are highly esteemed for bedding out and also for Winter bloom in pots; they flower very quickly and will soon fill your window-garden with beauty and fragrance. They are entirely hardy in open ground, require very little care, and produce great numbers of splendid flowers very early in Spring. The varieties we offer are the finest improved kinds and a few roots planted now will brighten up your flower beds and window-garden in the most surprising manner; they are all exceedingly beautiful.

The prices include postage, which is paid by us.

Large Trumpet Narcissus.

ARD REIGH—(Irish King.) Truly magnificent, large bold flowers, rich golden yellow trumpets, very early and fine, scarce and handsome. 10 cts. each, per doz., $1.00.

EMPRESS—A magnificent variety, large bold flowers, borne erect, and considered the finest of the two-colored trumpets. 10 cts. each, per doz., $1.00.

SIR WATKIN—The Giant Chalice Flower or Big Welchman. This is the largest daffodil grown, immense long-stemmed flowers, sometimes 5¼ in. across, rich lemon yellow, large dark cup tinted with orange, grand and handsome. 12 cts. each, per doz., $1.25.

GOLDEN SPUR—This is one of the grandest of daffodils, extra large, bold rich yellow flowers and broad handsome foliage. A strong grower and early bloomer, unsurpassed for garden or house culture. 10 cts. each, per doz., $1.00.

HORSEFIELDI—"The Queen of Daffodils." Extra large pure white flowers with rich yellow trumpets, very grand and beautiful, early and free bloomer, extra fine in every way. 10 cts. each, per doz., $1.00.

PRINCEPS—Very early and one of the best and most popular Winter-flowering varieties, largely grown for cut flowers, very large petals, soft sulphur yellow, with immense rich yellow trumpets. 4 cts. each, 6 for 20 cts., per doz., 40 cts.

TRUMPET MAJOR—Bright golden yellow, one of the best for bedding and fine for Winter flowering, an early and abundant bloomer. 4 cts. each, 6 for 20 cts., per doz., 35 cts.

GRANDEE—Flowers extra large, with broad creamy white petals and magnificent golden yellow trumpets, beautifully crimped and ruffled, exceedingly beautiful. 12 cts. each, 3 for 35 cts., per doz., $1.25.

Special Offer. Set of 8 Splendid Trumpet Narcissus, postpaid, only 65c.

Double Narcissus
or Improved Daffodils.

ALBA PLENA ODORATA—"The Double White Poet's Narcissus," or Gardenia-Flowered Daffodils. Double snow-white gardenia-like flowers, delightfully sweet scented. 3 cts. each, 3 for 8 cts., per dozen, 25 cts.

INCOMPARABLE—Fl. Pl. Large, handsome flowers, as double as roses, bright canary yellow, with rich orange centre. 3 cts. each, 3 for 8 cts., per doz., 25 cts.

ORANGE PHŒNIX—Outer petals nearly pure white, centre, mixed orange and white, very handsome. 5 cts. each, 6 for 25 cts., per doz., 50 cts.

VON SION A famous old kind, unsurpassed in beauty, single outside petals with long double trumpet, filled with beautifully crimped petals, extra fine for cutting. 4 cts. each, 3 for 12 cts., per doz., 35 cts.

The Four Varieties for 15 cts.

Narcissus Poeticus.

These beautiful varieties are entirely different from others; they have elegant star-shaped flowers with miniature saucer-shaped cups in the centre, they bloom quickly and very abundantly, and are equally valuable for garden planting and blooming in pots.

POETICUS ORNATUS—"The Improved Poeticus." Beautiful star-shaped flowers, pure white with saffron-colored cup tinged with rosy crimson, blooms quickly and is lovely for Winter bloom in pots, also for garden planting. 4 cts. each, 3 for 10 cts., per doz., 35 cts.

POETICUS—"The Pheasant's Eye or Poet's Narcissus " Pure white star-like flowers with orange cup, edged with crimson; hardy and fine for bedding, splendid for cut flowers. 2 for 5 cts., per doz., 20 cts.

C. & J. New Pedigree Cannas.

FOR WINTER BLOOM.

THE following **New American Pedigree Cannas** were all originated by us, and are among the very best kinds for house culture and Winter bloom.

NOTE:—More than fifty other magnificent varieties for Garden, Lawn and Park planting will be found fully described in our New Spring Guide for 1900. FREE TO ALL INTERESTED.

New Dwarf Canna, Golden Pearl.—Blooms finely in small pots when only six or seven inches high. The flowers are large and borne in handsome clusters of eight or nine at a time. The color is deep yellow with rich maroon centre. Very handsome. Nice pot plants, 25c., 3 for 60c., postpaid.

New Crimson Canna, Duke of Marlborough.—Dark rich velvety maroon, almost black, fine shapely flowers in large handsome trusses. Fine for pot culture. Nice pot plants, 25c. each, 3 for 60c., postpaid.

Martha Washington.—This is the grandest pink Canna yet introduced. Flowers very large and trusses immense. Pure bright rose pink. Pot plants, 30c., 3 for 75c., postpaid.

Buttercup.—Clear bright buttercup yellow, large, handsome flowers in fine open trusses, One of the most pleasing and attractive pure yellow varieties we have. Greatly admired by all. 35c. each, 3 for 90c.

New Gold-edged Canna, Gloriosa.—Superb flowers, centre of petals dark rich crimson, with border of deep golden yellow. An early and continuous bloomer, invaluable for house culture and bedding. Pot plants, 25c. each, 3 for 60c., postpaid.

Special Offer The set of 5 for **$1.25**, postpaid. **$1.00**, by express.

Ornithogalum Arabicum.

Arabian Star of Bethlehem—A really beautiful bulb for pots, large and solid like a hyacinth, and requiring the same treatment; flower stalks eighteen to twenty inches high, bearing immense clusters of large, pearly white flowers with jet black center, rich aromatic fragrance. The flowers last a long time and are almost unrivalled for beauty and fragrance. They do nicely in open ground if protected from hard freezing. 5 cts. each, 6 for 20 cts., per doz., 40 cts.

Ornithogalum Arabicum.

Narcissus Corbulata or Bulbocodium.

These are curious and very pretty varieties, both for winter bloom in pots, and also for planting in open ground. Large trumpet bell-shaped flowers, bright golden yellow, pale sulphur and cream white, delightfully fragrant and very beautifull. They bloom finely in February and March and are a charming addition to the window garden. Plant several together in a pot or box.

Bulbocodium.—"The Large Yellow Hoop Petticoat," Rich golden yellow. 10 cts. each, $1.00 per dozen.

Citrinus.—"The Large Sulphur Hoop Petticoat," large pale sulphur trumpets, very beautiful. 5 cts. each, 50 cts, per dozen.

Algerian.—"White Hoop Petticoat," pure snow white, very early. Will bloom at Christmas if planted in September. 5 cts. each. 50 cents per doz. **The 3 Varieties for 18c. Mixed Colors 2 for 10c. 50c. per doz.**

Narcissus Corbulata.

SCILLA CLUSI.

Scilla Clusi (the Peruvian Hyacinth or Cuban Lily)—A very grand and striking plant for house culture, throws up a strong bold flower stem, with long lance-shaped leaves, and bearing one enormous pyramidal cluster of rich bright blue star-shaped flowers. It is very easily grown, commences to bloom about mid winter and continues a long time. When warm weather comes, plant in open ground, and if taken in before freezing, it will be ready to bloom again next winter. 15 cts. each ; 2 for 25 cts.; $1.50 per doz.

Erythronium Grandiflora.

The Beautiful Wood Lily.

These pretty bulbs send up five or six blooms each, large, drooping, lily-like flowers, rich canary yellow, and borne on long slender stems. The foliage is very handsome, as well as the flowers, they make highly ornamental pot plants, and when planted in open ground, will soon take care of themselves and bloom beautifully every spring. 4 cts. each ; 3 for 10 cts.; 12 for 35 cts.

JONQUILS.

The jonquils are lovely low growing little flowers, and like the hyacinths and crocus, bloom very early in spring and are very pretty and attractive. They are bright golden yellow, and delightfully sweet scented. Highly valued for bedding and borders, and also for house culture in pots or pans, several should be planted together for best effect.

Scilla Clusi.

Double Jonquils—Extra fine deep yellow. 5 cts. each, 3 for 15 cts.; 50 cts. per doz.

Single Jonquils—Rich yellow, very fragrant. 3 for 5 cts.; 6 for 10 cts ; 20 cts. per doz.

Giant Campernelle (Rugul sus)—Twice the size of the others and same beautiful bright golden yellow, deliciously sweet-scented. Extra fine for both pot culture and bed ing out. An early and profuse bloomer. You want it. 4 cts. each ; 3 for 10 cts.; 35 cts. per doz.

Campernelle Mixed—Large yellow flowers, rich and handsome. 3 for 5 cts ; 20 cts. per doz.

Tritelia Uniflora.

Spring Starflower, One of the most charming fairy-like white flowers for Winter. White star like flowers tipped and faintly striped with blue. Planted several together in-doors, they bloom finely for months, also fine for bedding and edging in open ground, forming a mass of starry blooms from early spring till summer. 3 for 5 cts.; 15 cts. per doz.

Blue Tritelia—Exactly like the above except that flowers are lovely violet blue, they contrast finely with each other, and should be planted together when convenient. Equally valuable for pots and garden. 3 for 10 cts.; 25 cts. per doz.

Tritelia Uniflora.

Fritillaria Meleagris (Guinea Hen Flower)

These new Fritillarias are remarkably beautiful and satisfactory bulbs, both for pot culture and open ground. The plant is neat and handsome, with bold and striking flowers of most brilliant colors, curiously checkered and variegated in distinct patterns; they are entirely hardy and succeed well everywhere. We recommend our friends to plant them liberally, both in pots and for bedding out.

Fritillaria Meleagris—Finest mixed, 2 for 5 cts., 4 for 10 cts., 12 for 20 cts.

Trillium Sessile—A fine hardy perennial, with handsome mottled foliage, and large, pure white, lily-like flowers; very handsome and desirable for a partially shaded place, where it will form a permanent bed, and increase in size and beauty from year to year. 10c. each, 3 for 25c., $1.00 per doz.

Babiana—These little bulbs are beautiful for winter bloom, five or six in a medium size pot make a lovely parlor ornament. They grow six to nine inches high, throwing up charming spikes of nodding, graceful flowers; rich self colors, ranging from deep crimson to brightest blue.

Fine Mixed Colors, 3 for 12 cts.,
per doz., - - - - **35c.**

The Calochortus...

Called also the Butterfly Tulip or Mariposa Lily, is a lovely bulbous flower, offered this year in our Blue Ribbon Window Garden, and shown in our colored plate on back cover. A half dozen bulbs can be set in a medium size pot; they will bloom quickly, and the colors are marvelously beautiful, real butterfly shades and markings; there are two kinds, the large- and small-flowered, both very beautiful.

Splendid Mixed Calochortus—All colors, large and small, 4 cts. each, 3 for 10 cts, 12 for 30 cts.

Christmas Rose (Helleborus Niger)—A beautiful hardy bulbous plant, blooming finely in all ordinary situations, when planted in pots in the house it blooms in December and continues through the winter, and if set in open ground will bloom profusely for weeks in the early spring. The flowers are two to three inches in diameter, and pure waxy white, quite rare and not often seen. Strong blooming bulbs, 30 cts. each, $3.00 per doz., postpaid.

Tritonia

When making up your order don't fail to include a few of the lovely New Tritonins, they are quite new and fresh and greatly admired; fine graceful sprays of soft, richly colored flowers, ranging from white, through salmon, orange and scarlet.
Fine mixed colors, 3 for 10 cts., 25 cts. per dozen.

BRODIA COCCINEA.

Brodia Coccinea

Floral Firecracker Plant

This is a particularly nice house-plant, and entirely different from everything else; treat the same as other house bulbs. The stems bear pretty clusters of long tubular-shaped flowers, deep rich crimson, tipped with white, very pretty and interesting. Price 5 cts. each, 3 for 12 cts., 40 cts. per dozen.

True Chinese Sacred Lily.

Chinese Sacred Lily.

The Great Chinese Sacred Lily is one of the most popular winter-flowering bulbs; vast quantities are brought from China every year. The bulbs are very large and each one sends up from five to twelve flower stems, which bear great clusters of large pure white wax-like blossoms, with yellow centres, and of a rich delicious fragrance. About one bulb in three will produce double flowers, the others single, as shown in the engraving. They grow nicely in pots of ordinary soil, but the best and quickest way to bloom them is in water. Fill a bowl or some similar vessel with pebbles or small stones, in which place the bulb, setting it about one-half its depth, so that it will be held firmly, then fill with water to the top of the pebbles, and place in a warm sunny window. The bulb will at once commence a rapid growth and bloom in two or three weeks. Ours are the genuine Chinese Sacred Lily, grown in China. Cheaper bulbs are offered, but they are an inferior grade, and not to be compared with these in beauty. **Ours are the genuine variety, grown in China. Price, extra large quick blooming bulbs, by mail, postpaid, 12 cts. each; 3 for 35 cts.; $1.35 per doz.**

Golden Sweet-scented **SACRED LILY**

(See full description, page 29). Grows in water and pebbles exactly like the Chinese variety, flowers bright clear golden yellow, and borne in the greatest profusion. Price, 4 cts. each; 3 for 10 cts.; 30 cts. per doz.

SPECIAL OFFER We will send one Chinese Sacred Lily, three Golden Sweet-scented Lilies, and two Poeticus Ornatus, the six only 25 cts., postpaid.

Allamanda Williamsii

Allamanda Williamsii.

ALLAMANDA WILLIAMSII.—This is a charming new house plant, very distinct and entirely different from other plants of this class; it makes a neat, compact bush, with dark, glossy green leaves, and bears grand trusses of large golden-yellow, lily-like flowers; very beautiful and deliciously sweet-scented. It is a very rare and handsome plant, and an elegant ornament for conservatory and house decoration. Price, strong plants that will soon bloom, 20 cts., larger size, 35 cts. postpaid.

NEW CUPHEA, LITTLE PET.—An elegant window plant; grows only eight or ten inches high; almost as round as a ball; fine, deep green leaves, and dotted all over with pretty rosy-pink flowers; blooms all the time. 10 cts.

NEW HELIOTROPE. Jersey Beauty.—Rich purple flowers, fine Winter bloomer, for living room or window garden. 10c.

LEMON VERBENA.—Highly prized for the spicy and delicious fragrance of its leaves, entirely different from anything else; nice for pots and bedding. 10 cts. each, postpaid.

PARLOR IVY.—A lovely climber for house or window; quick grower, makes an abundance of pretty green foliage. 10 cts. each, postpaid.

SANSEVERIA ZEALANICA.—A beautiful house plant. The leaves grow erect, one to two feet high, and are beautifully barred, white and green; always showy and attractive, and requires very little care. 15 cts. each.

TRADESCANTIA VARIEGATA.—Fine variegated leaves, crimson and green; excellent for vases, etc. 10 cts.

THE LAST SIX, ONE EACH, ONLY 35c., POSTPAID.

WINTER-BLOOMING FREESIAS.

FREESIA REFRACTA ALBA. This is the True Large-flowering Freesia; the flowers are pure snow-white, delicately tinted with pale yellow and very sweet. Next to Violets, they are the sweetest Winter-blooming flowers we have, and are extensively planted by the **Hundred** and **Thousand** for cutting. They retain their beauty for a long time, and by planting at different times, you can have a succession of lovely bloom all Winter. Freesias are very easily grown, a dozen bulbs can be planted in one good sized pot or pan of ordinary soil; water sparingly and set in a warm sunny window, and they will soon begin to bloom. We send best quality bulbs, all sure to bloom, at the prices here given.

Freesia Refracta Alba, extra large selected bulbs, 3 for 10c., 25c. per doz., $1.25 per 100, postpaid; $10.00 per 1000. First size bulbs; all sure to bloom, 12c. per doz., 75c. per 100; $6.00 per 1000.

New Large-flowering Hybrid Freesias.
This is a grand new strain with flowers of largest size and many beautiful colors not seen before, including white, yellow, pink, purple, violet and rose; delightfully fragrant. **Price, all colors mixed, 4c. each, 3 for 10c., 25c. per doz., $1.25 per 100.**

Scarlet Freesia.
Anomatheca Cruenta. This resembles the White Freesia in all respects, except that its flowers are bright scarlet, shaded crimson, and though it does not flower quite so early, it is very bright and handsome, and makes a beautiful contrast planted with the others. **Price, 4c. each, 3 for 10c., 30c. per doz., 1.75 per 100, postpaid.**

Large Yellow Freesia.
Leichtlinii Major. A very pleasing variety and makes a charming combination planted with the others. Large orange yellow and white flowers, very fragrant and beautiful. 4c., 3 for 10c., 30c. per doz., 2.00 per 100, postpaid.

SPECIAL OFFER. 6 White, 3 New Hybrid, 3 Scarlet and 3 Yellow, 15 in all, for **25** c. postpaid.

New Hardy Gladiolus, The Bride.

These are very fine for Winter-flowering in the house, six bulbs can be planted in one six-inch pot. They begin to bloom very quickly, the flowers are large pure snow white, and borne in long slender spikes, very beautiful and largely used for cut flowers. The Bride Gladiolus is also quite hardy, and when planted in the flower bed will bloom splendidly in the Spring. It is equally desirable both for Winter bloom in the house, and planting in the flower bed. Fine blooming bulbs, 3 for 10c., 25c. per doz., $1.35 per 100, postpaid.

GIANT CYCLAMEN.

Cyclamens are among the most charming bulbs we have for parlor and greenhouse culture. They are very easily grown, bloom freely, and continue a long time in bloom. The flowers are exceedingly beautiful, and range through many shades of pink, crimson and white. The leaves are finely variegated, so that the whole plant is highly ornamental, and particularly suitable for house and conservatory decoration. They like lots of light and air, but not too much heat. All lovers of beautiful house plants should by all means have some pots of Cyclamen. When done blooming the bulbs should be dried off and allowed to rest like Callas.

CYCLAMEN GIANT. (True.)—First size dry bulbs, finest colors mixed. 20 cts. each, 3 for 50 cts., $2.00 per doz., postpaid.
CYCLAMEN PERSICUM.—Similar to above in every way, but flowers not so large. 15 cts. each, 2 for 25 cts., $1.50 per doz., postpaid.

Cyclamen.

THE BEAUTIFUL JPANIJH IRIS.

These are fine for blooming in pots in winter; they bear lovely large orchid-like flowers of most brilliant colors, blue, purple, yellow, pearly white and black, beautifully variegated, striped, spotted and ruffled; they are also entirely hardy and splendid for bedding in open ground, need no protection and will bloom finely every spring without attention. We put the price very low and hope all who can will give them a liberal trial.

The Finest Named Varieties

Peacock—Pure white, with a bright blue spot on each petal, very showy and handsome. 5 cts. each; 6 for 25 cts.; 50 cts. per doz.

The Queen—Large beautiful pure white flowers. 3 cts. each; 4 for 10 cts.; 25 cents per doz.

Belle Chinoise—Deep golden yellow, very fine. 3 cts. each; 4 for 10 cts.; 25 cts. per doz.

Olympia—Azure blue and royal purple with yellow markings. 3 cts. each; 4 for 10 cts.; 25 c.s, per doz.

Prince of Orange—Orange yellow, with rich purple spots. 3 cts. each; 4 for 10 cts.; 25 cents per doz.

SET OF FIVE NAMED VARIETIES FOR 12 CENTS.

Spanish Iris Finest colors mixed. 3 for 5 cts., 15 cts. per doz.; $1.00 per hundred.

Spanish Iris.

NEW PALESTINE IRIS.

We import these magnificent Iris direct from Palestine, and they are probably the most interesting novelty for house and winter bloom recently offered ; the flowers are large and of very remarkable form, and rival the finest orchids in beauty and variety of colorings and markings. We recommend them especially for house culture, as they grow and bloom nicely in pots without special care, but would scarcely stand the winter in open ground unless carefully protected. They are one of the choicest and most interesting novelties you can have for winter bloom. Stock is limited and orders will be filled strictly in rotation.

ATROPURPUREA (The Black Iris)—Extra large and beautifully shaped flowers of deep dark, almost black color, fine sutiny texture and lustre; a rare and very attractive flower. Price, 20 cts. each. Dry bulbs only.

HISTRIO—Several beautiful colors are seen in this lovely flower, deep sky blue, exquisitely variegated with various shades of white and golden bronze; besides being a flower of rare and unique beauty, it is delightfully perfumed with a rare fragrance exclusively its own. Dry bulbs, 20c. each.

MARIAE—Beautiful flowers, true fleur-de-lis form, color lovely bright lavender pink, exquisitely shaded and clouded with violet purple, 20 cts. each.

SARL NAZERENAE—This noble variety is considered one of the most beautiful of all, flowers large and of fine substance, ground color pale canary, elegantly blotched and shaded with crimson and brown, and veined with blue. Indescribably beautiful, 20 cts. each.

The 4 Varieties for 65 cts., Stock Limited.

Ixias and Sparixias.

The **Ixias and Sparixias** are house bulbs, not hardy in open ground, but extra fine for pots, as they are sure to grow and produce an abundance of splendid flowers of most brilliant and striking colors. Though somewhat alike in growth the colors and form of flowers are entirely different, so that they make an elegant contrast planted together, six or a dozen in a pot or pan. They bloom quickly and are always admired. Price the same for both. 2 for 5 cts.; 20 cts per doz.

Anemone, Coronaria—Large double flowers, almost like hollyhocks, two or three brilliant colors blended in each flower. Fine mixed. 3 for 10 cts.; 25 cts per doz.

Anemone (Single Scarlet)—Rich, dazzling flowers, very bright and striking, splendid for pot culture, also for open ground. 3 for 12 cents; 30 cents per dozen.

Ixias and Sparixias.

Anemone, Apennina—Beautiful large rich blue flowers and elegant cut foliage, blooms profusely early in spring, also fine for pots. 3 for 10 cts.; 25 cts. per dozen.

CROCUS..

Finest Named Large Flowering Varieties

ALL POSTAGE IS PAID BY US.

FIRST FLOWERS of SPRING.

COPYRIGHTED 1893

CROCUS.

AMONG all the pretty flowers of Spring, none are greeted with a heartier welcome than the Crocus. Their flowers are lovely in form, and of the brightest, freshest colors. Pure White, Golden Yellow, Purple, Pink and Variegated. They are entirely hardy, and like a rich dry soil and sunny location. They are suitable for planting in beds and borders, also for edging in ribbon lines of any desired color, also for planting in the grass in lawn or yard. CROCUS ARE ALSO VERY HIGHLY VALUED FOR HOUSE CULTURE. They bloom very quickly and are so cheap that all can afford to have them. OURS ARE THE IMPROVED STRAINS, NOTED FOR LARGE FLOWERS AND BRIGHTEST COLORS.

CROCUS MAMMOTH YELLOW.

This is one of the grandest and most beautiful varieties of all, the bulbs are of mammoth size, and each one throws up from eight to twelve extra large flowers of the brightest and richest golden yellow. It is very early and exceedingly showy and handsome. A few bulbs planted in a five or six-inch pot will surprise you with their beauty and profusion of bloom, and by planting at different times a succession of lovely golden flowers can be had all Winter. A bed planted in the open ground is always an object of striking beauty very early in Spring, and if planted in lines on a well-kept lawn it is easy to write any desired name in letters of gold in the green grass. **Price, large bulbs, 3 for 5c., 12 for 15c., $1.00 per 100, postpaid.**

The FINEST NAMED CROCUS.

Baron Von Brunow—Dark violet blue, a grand early sort. Price, 12 cts. per doz., 75 cts. per 100.

Cloth of Gold—A beautiful variety from the far East, one of the earliest Spring-flowering kinds, handsome feathered flowers, golden orange, tinged with purple and scarlet. Price, 12 cts. per doz., 75 cts. per 100.

Cloth of Silver—Same as above, except the color, which is lovely silvery white. Price, 12 cts. per doz., 75 cts. per 100.

Sir Walter Scott—White, striped purple, very large. Price, 12 cts. per doz., 75 cts. per 100

Prince of Wales—Very large, dark purple. Price, 12 cts. per doz., 75 cts. per 100.

Set of Six Varieties, three each, eighteen in all, for **15 cts.**

Best Varieties Mixed White and light, 10c. per doz., 50c. per 100; Purple and Blue, 10c. per doz., 50c. per 100; Striped and Variegated, 10c. per doz., 50c. per 100; Yellow and Orange, 10c. per doz., 50c, per 100; All Colors Mixed, 10c. per doz., 45c. per 100.

Snow Drops Galanthus.

Snow Drops

GALANTHUS.

Snow Drops are the earliest of all spring flowers, as they bloom a few days before the Crocus and Scilla Their drooping white blossoms are exceedingly pretty and graceful, and make a beautiful contrast planted with the lovely blue Scilla in beds or in the grass on the lawn (See page 42.)

Single Snow Drops—3 for 5 cts.; 15 cts per doz. ; $1.00 per 100.

Double Snow Drops—Flowers perfectly double and pure snow white. 4 for 10 cts.; 25 cts. per doz., $1.75 per 100.

Giant Snow Drops—(Elwes). The largest and most beautiful of all snowdrops, flowers pure snow white and deliciously sweet, extra fine for pots and bedding. 2 for 5 cts.; 15 cts per doz.; $1 per 100.

Glory of the Snow— (Lucillae). A lovely variety producing large clusters of beautiful azure blue flowers with clear white centre, very handsome and desirable. 2 for 5 cts.; 20 cts. per doz.

Giant Chionodoxa—Much larger flowers than the above, lovely violet blue with white eye, very striking and beautiful. 2 for 5 cts.; 20 cts. per doz.

...ALLIUM...

Azureum—The Blue Allium, a rare and beautiful variety not often seen, throws up a profusion of lovely sky blue flowers, star shaped and borne in pretty graceful sprays, extra fine for pots and also for bedding. Price, 10 cts. each ; 3 for 25 cts ; 75 cts. per doz.

Aureum—(Moly or Golden Allium.) This is a fine hardy garden plant, very desirable for beds and borders, bears large golden yellow flowers in June, and is very showy and handsome, but not as well known as it ought to be. Grows one foot high. Price, 3 for 10 cts.; 25 cts. per doz.

Allium Neapolitanum—This is the most popular of all the Alliums, and especially valuable for pot culture. Grows very easily, costs but a trifle and bears great masses of lovely pearly white flowers, lasting a long time. Five or six bulbs in a four-inch pot make a beautiful window ornament. Hardy also in open ground, fine for bedding and borders, and always greatly admired, whether in garden or pots. Price, 2 for 5 cts.; 25 cts. per doz ; $1.25 per 100, postpaid.

Allium Neapolitanum.

Special Offer. 1 Blue Allium, 3 Golden Yellow, 3 white; 7 for 20 cts., postpaid.

RANUNCULUS—Fair Maids of France.

Sturdy dwarf-growing plants, highly valued for pot culture and for bedding out. In the North they should have a covering of leaves or litter during Winter. They bloom quite early and will produce an abundance of large double brilliant colored flowers.

Double Turban Mixed—Peony formed flowers, large and early. Many brilliant colors. 2 for 5 cts.; 20 cts per doz.

Double Giant French Mixed—Strong vigorous growers, bearing immense gorgeous flowers. 2 for 5 cts., 20 cts. per doz.

Double Persian Mixed—Rose shaped flowers, very double, rich handsome colors. 2 for 5 cts.; 20 cts. per doz.

Lily of the Valley—Its beautiful sprays of lovely pure white bell-shaped flowers are always greatly admired. It is perfectly hardy, does not object to shade, and will do well in any odd corner, and increase in number and beauty from year to year. In doors, a half dozen bulbs in a four-inch pot, kept cool awhile, and then given light and moisture, will soon produce an abundance of lovely fragrant flowers. Price, strong flowering crowns, by mail, postpaid, 5 cts. each ; 6 for 25 cts.; 12 for 50 cts. *Ready about Nov. 15th.*

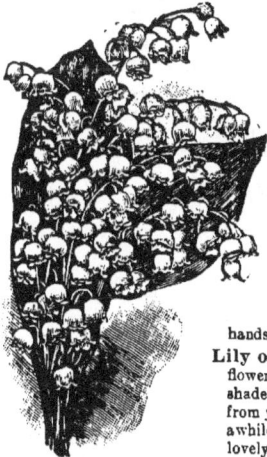

Lily of the Valley.

Important ★ Announcement
A BOON FOR   OUR PATRONS

FLOWERS are Nature's Sweetest Gift to Man. They are loved and admired by all, but not everyone has success in growing them. Our business is the sale of plants, seeds and bulbs, but our interest in your purchase does not end when your order has been filled. We want the flowers you get of us to grow and bloom, and we are trying to give you all the help we can to make them do so. It is with great pleasure, therefore, that we announce to our customers the completion of arrangements which will enable us to give, FREE, with orders sent us this Fall, a subscription to the best, brightest and most practical floral journal in existence.

"HOW TO GROW FLOWERS"

is the title of this magazine, and no publication was ever more aptly named. It is an up-to-date Encyclopedia of Floral Information, which no lover of flowers can afford to do without. It tells just what you want to know about your plants; what attention they need, and when and how it should be given. The best authorities on such subjects, (Eben E. Rexford and others), are regular contributors, and tell JUST HOW THEY have been able to make flowers thrive and bloom. No theories, but every day plain and practical information; 36 large pages, beautifully illustrated. It also has a department: "Things to do in the Month," another, "Answers to Correspondents." It is in short your IDEAL FLORAL MAGAZINE.

OUR UNPRECEDENTED OFFER
(READ IT WITH CARE)

We want all our customers to enjoy the great assistance and pleasure that this Magazine brings with it each month; and we are glad to make this remarkable offer. It is as follows: To every person not already a subscriber to "HOW TO GROW FLOWERS" who orders Plants, Bulbs, or Seeds from this Catalog to the value of 50 cts. or more, on or before December 1st, 1899, we shall present, ABSOLUTELY FREE, as a premium, on request, a six months subscription to this Magazine, beginning with September issue.

REGULAR PRICE. of this magazine for one year is 50 cents.	OUR OFFER. 6 months with every 50c. order FREE OF CHARGE.	CALL YOUR NEIGHBORS' attention to this matchless offer, and get them to order with you. THEY WILL THANK YOU FOR IT.

Remember THE CONDITIONS THAT WILL ENTITLE YOU TO A SIX MONTHS SUBSCRIPTION FREE OF CHARGE.

You must be a person who is not already a subscriber. Your order from this catalog must amount to 50c. or more. It must be received by us on or before December 1st, 1899.

The Biggest Dollar Flower Garden Ever Offered.
20 Fine House Plants, Specially Prepared for Winter Bloom, only $1.00, Postpaid.

1 Washington Weeping Palm, Filefera.
1 New Double-flowering Nasturtium, Sunset. (Page 13.)
1 New Flowering Begonia, Marguerite.
1 New Flowering Begonia, Selected.
1 House Rose, Yellow Soupert.
1 House Rose, New Red Pet.
1 New Abutilon Savitzii. Page 18
1 New Justicia Velutina.
1 Tradescantia, Joseph's Coat.
1 Red Winter-blooming Geranium.
1 New Ageratum, Princess Pauline, (Page 42)

1 Pure White Geranium.
1 Parlor Ivy, Elegant Parlor Vine.
1 Cuphea, Little Pet. Blooms all the time.
1 Lemon Verbena. Fragrant Leaves.
1 Pretty House Fern. Fine for Pots and Baskets.
1 New Weeping Lantana, Mrs. McKinley.
1 Fuchsia Speciosa. True Winter Bloomer.
1 Fuchsia, Jupiter. Grand Flowers.
1 Swansonia Rosea. New Rose Colored.
1 Heliotrope, Jersey Beauty. Very sweet

Presented FREE with Every Order.

21 in all, only $1.00, Postpaid.

Six months subscription to the New Floral Magazine, "HOW TO GROW FLOWERS," FREE on request, with this collection.

NEW GOLDEN SACRED LILY.

Golden Sacred Lily.

A new and splendid bulb for house culture ; belongs to the same family as the Chinese Sacred Lily, but the flowers are much larger, exceedingly sweet, and the color is clear **Bright Golden Yellow.** Each bulb produces several spikes of bloom, and will thrive in pots of soil or sand, or in a bowl of pebbles and water like the Chinese kind. Bulbs flower very quickly and may be had in bloom for Christmas or New Years, or even earlier. Hard usage or even freezing will not kill them, or keep them from blooming. **They are extra hardy, fragrant and very beautiful ; all should give them a trial.** Three or four in a five-inch pot make an elegant window ornament, and when planted in the garden will bloom almost as early as the Crocus. **Price, 4c. each, 3 for 10c., per doz., 30c.**

SIX BEAUTIFUL SACRED LILIES.

1 GREAT CHINESE SACRED LILY—Largest Size Bulb.
3 NEW GOLDEN   "   "   Clear Bright Yellow.
2 DOUBLE ROMAN   "   "   Pure Snow White.

The Six, postpaid, only **25 cts.**

Winter-blooming Oxalis

Bermuda Butter Cup Oxalis.

These beautiful Oxalis are the **true winter-blooming kinds.** Absolutely unequaled for pots, baskets, vases, window boxes, etc. They begin blooming very quickly and continue to throw out their lovely buds and blossoms every day all Winter Nothing finer for window culture.

Bermuda Butter Cup Oxalis.—One of the finest Winter-blooming plants ever seen, a strong vigorous grower with handsome foliage, and bearing a constant succession of lovely yellow flowers all Winter. Well-grown plants have produced as high as seventy flower stems at one time, and over one thousand flowers in a season. 4 cts. each, 3 for 10 cts., 35 cts. per doz.

Oxalis Boweii.—Large flowers, bright rich pink. 4 for 10 cts., 12 for 25 cts.

Snow Ball. (Multiflora Alba.)—Large pure white flowers. 4 for 10 cts., 25 cts. per doz.

Double Yellow Oxalis.—Large double flowers, bright rich yellow. 4 cts. each, 3 for 10 cts., 35 cts. per doz.

Versicolor.—Deep rose with white center, beautiful. 3c. each, 4 for 10c., 25c. per doz.

THE SIX NAMED KINDS, ONE EACH, 15 cts.
TWO EACH, TWELVE IN ALL, FOR 25 cts.

Fine Mixed Oxalis—10 cents per dozen; $1.00 per hundred.

BIZARD AND BYBLOOM TULIPS.

We ask special attention to the Bizard, Bybloom and Breeder types of Tulips. Their large, cup-shaped flowers somewhat resemble the Gesneriana type, and are always greatly admired for their brilliant and striking colors The **Bizard** have yellow grounds, flamed and blazed maroon, black, scarlet, bronze and brown, The **Byblooms** have white, light or violet grounds, flaked and feathered with rose, pink, purple, scarlet, black and crimson, elegantly variegated. Price, fine mixed **Bizard Tulips,** 3 cts each, 25 cts. per doz., postpaid. **Bybloom Tulips, fine** mixed, white grounds, elegantly variegated. 3 cts. each, per doz., 25 cts. **Bizard and Bybloom Tulips,** all colors mixed. 3 cts. each, per doz., 25 cts.

New Dwarf Justicia Velutina

A grand new plant for pot culture and bedding. Our customers write: "It bears a surprising number of lovely pink plume-shaped flowers, and gives great satisfaction." Treated like a geranium, it will continue blooming as long as kept in growing condition, and when planted out in the spring, will bloom the whole season.

15c. each, 2 for 25c., larger plants, 25c. each, exp.

New Dwarf Justicia Velutina

New Flowering Begonias

NEW FLOWERING BEGONIA, BIJOU—This is an elegant new variety for parlor or conservatory, makes a handsome compact plant of true sugar-loaf form; clothed all over with round glossy green leaves, and glistening rose-pink flowers. Easily grown, likes partial shade, and is a constant and profuse bloomer; highly valued for house culture. 15 cts. each, 2 for 25 cts., $1.50 per doz.

LOBATA VARIEGATED—Long pointed leaves, prettily blotched with silver spots, rosy white flowers, very handsome. 10 cts. each.

MARGUERITE—Fine bronzy green foliage, curiously veined, neat, bushy form, large clusters of lovely colored flowers. 10 cts. each.

THURSTONII—Leaves bronze-shaded crimson and red, rosy-white flowers, large clusters, handsome. 10 cts.

SANGUINEA—Large thick leaves, upper side rich olive, under side deep crimson, lovely rosy-white flowers in long racemes. 10 cts. each.

SAUNDERSONII—Low compact grower, with glossy green leaves, and large clusters of scarlet heart-shaped flowers, beautiful. 10 cts. each.

VERNON—Immense bloomer, lovely blush and pink flowers. 10 cts. each.

Set of Seven for 50 cts., postpaid.

Rex or Painted Leaf Begonias

QUEEN VICTORIA—Rich, velvety leaves, covered with red plush effect, edged with green and dotted with silver. 15 cts. each.

SPECULATA—A fine plant, with leaves like a grape leaf. They are bright green in color with a background of chocolate; the veins are of a light pea-green, and the whole leaf spotted with silver. 15 cts.

Royal Purple Bougainvillea. (New)

This is an immense winter bloomer, and continues for months at a time, covered all over with its gorgeous clusters of royal purple blossoms; one of the handsomest winter-blooming flowers ever introduced. Very beautiful and entirely different from everything else in color and form. Red out in summer. Can be kept for years like an Otaheite Orange or Oleander. A constant and abundant bloomer. Very desirable and handsome. Blooming size plants, 15 and 20 cts. each, postpaid. Extra size three year plants, two and one-half feet, $3.00 each, by express.

Royal Purple Bougainvillea.

Gloxinia — GRAND ERECT-FLOWERING VARIETIES

(Dry bulbs in separate colors ready in November.)

The Gloxinias are known as among the choicest and most beautiful flowering plants we have for Winter and Spring bloom in living-room, conservatory or greenhouse. They grow easily from the bulbs, and bloom finely for several months; each bulb will make a large plant, and produce many splendid flowers, three to four inches across, and of the most gorgeous and exquisite colors; some are spotted and mottled, others beautifully variegated, and all indescribably rich and velvety. They are recognized by all as entirely out of the usual, and among the most rare and handsome house plants to be had. When done blooming, the bulbs should be dried off and laid away till time to plant again. We offer three colors: scarlet, blue and white. 15 cts. each, 3 for 40 cts., $1.50 per dozen, postpaid. Good blooming bulbs, second size, 10 cts. each, 3 for 25 cts., $1.00 per dozen, postpaid.

Gloxinia

New Perpetual Blooming Sweet Violets.

OUR Violets are remarkably healthy and vigorous, and we offer the following choice varieties in two sizes: **Fine Pot-grown Plants for Mail Orders, and Strong Field-Grown Clumps for Express Orders.** Both are strong, well-established plants, which will begin to bloom very quickly, and should bear a large crop the present season. Violets like plenty of light and air and moderate moisture, but not too much heat.

POT PLANTS, 10c. 85c. per doz. postpaid.
FIELD CLUMPS, Express only, 15c. $1.50 per dozen.

LUXONNE. (New.)—Single flowers of very largest size, rich violet purple, deliciously fragrant, free bloomer and entirely hardy. 10 and 15c. each.

PRINCESS OF WALES. (New.)—Extra large flowers, tremendous bloomer, clear deep blue, very fragrant and vigorous, entirely hardy. 10 and 15c. each.

RUSSIAN.—Splendid large single flowers, deep rich blue, very sweet, entirely hardy, very productive. 10 and 15c. each.

MARIE LOUISE.—Large double flowers, deep violet purple, very sweet and productive. 10 and 15c. each.

Russian Violets.

SHONBRUN.—Beautiful dark blue flowers, borne in wonderful profusion, deliciously fragrant and entirely hardy. 10 and 15c. each.

SWANLEY WHITE.—The finest white violet in cultivation, profuse perpetual bloomer, perfectly double and exquisitely fragrant. 10 and 15c.

PRICE, NICE POT PLANTS, 10c. each, 3 for 25c.; 85c. per dozen, postpaid. FIELD CLUMPS, by express only, 15c. each, 6 for 75c., $1.50 per doz.

The Best Hardy Lilies.

THESE beautiful lilies are quite hardy, but should be planted four or five inches deep, and given a light covering of leaves or litter before the ground is deeply frozen. They should not be disturbed, but left to grow on from year to year. They get larger and finer with age.

LILY AURATUM—The gold banded lily of Japan, considered the Queen of Lilies, and the most beautiful of all immense flowers, nearly a foot in width, borne in great clusters, seeming more than the slender stem can bear; color, rich creamy white, thickly spotted with crimson and brown, each petal having a wide golden yellow band through the center; very fragrant and sure to bloom. First size good blooming bulbs, 15c. each, 2 for 25c., per doz., $1.35. Extra size bulbs, 20c. each, 3 for 50c., doz.; $1.85, postpaid.

Lilium Pardalinum—The California Leopard Lily. An elegant and very beautiful lily from California. Rich scarlet and yellow flowers, spotted with purplish brown. 15c. each, $1.50 per doz.

Canadense—One of our finest lilies, bearing graceful clusters of drooping bell-shaped red and yellow flowers. 10c., $1.00 per doz.

L. Superbum—Stands at the head of our native lilies; flowers bright orange red thickly spotted with purple. 15c , 2 for 25c.

L. Album—Extra large flowers, pure snow white, very sweet scented. 18c. each, 2 for 30c.

Batemanii—Pure apricot color, without spots. A highly prized Japanese variety. 15c. each, 2 for 25c.

L. Elegans—Fine, crimson and dark red. 15c. each, 2 for 25c.

L. Longiflorum—Beautiful long trumpet-shaped flowers, pure snow white, very fragrant and entirely hardy. 15c. each, 2 for 25c.

Madonna Lily—Pure snow white, very handsome and richly scented. 12c. each, $1.25 per doz.

L. Melpomene (Speciosum)—Rich blood red with a clear frosty white border; very handsome and hardy. 20c. each.

L. Roseum—One of the very best kinds, splendid large flowers, rose and white, spotted crimson; beautiful. 15c.; $1.50 per doz.

L. Tigrinum—Fl. Pl. Extra large double flowers, bright rich orange, spotted black. 15c. each, $1.50 per doz.

L. Wallaceii—Beautiful Japan Lily; clear buff, elegantly spotted with crimson; very handsome and desirable. 15c. each.

Special Offer Two 15c. kinds, 25c., six 15c. kinds, 75c , complete set, 13, postpaid, for - - - - - **$1.50**

Lily Auratum,

C. & J. Little Gem Calla Lily.

New Dwarf Calla Lily...

Little Gem

MOST remarkable Calla Lily ever introduced. Grows only half as tall as the old kind, and bears twice as many flowers.

We send Extra Fine, Imported Dormant Bulbs. The very best for quick and abundant bloom, 15c. each, 2 for 25c., $1.50 per doz.

New Coleus...

..Dr. Ross

MOST splendid Coleus ever seen, enormous leaves, rich dark velvety crimson, with a broad wedge of creamy white, beautifully fringed and bordered with gold. An elegant house plant. 15c. 4 for 50c.

The Wonderful BLACK CALLA

(ARUM SANCTUM.)

This wonderful variety resembles the White Calla in growth and foliage, but the flowers are rich dark purple, and the spikes or spadix is coal black, very curious and remarkable and always attracts a great deal of attention, especially when known that it is a native of Palestine, and has only recently been brought to this country from the neighborhood of Jerusalem. Nice, dry bulbs, ready for potting. Extra Size Bulbs, 20 cts. each, 2 for 35 cts., $2.00 per doz. First Size Bulbs, 15 cts. each, 2 for 25 cts., $1.50 per doz.

The White Calla or Lily of the Nile

Well-known as one of our most valuable plants for house culture, we send extra fine dormant bulbs, ready to pot immediately; they like good rich soil and plenty of heat and moisture and will bloom finely, very quickly—5- and 6-inch pots will be right for the two sizes—when done blooming they should be dried off and allowed to rest three or four months before starting up again. Dry bulbs only.

Extra Large Bulbs, 20 cts. each, 2 for 35 cts. First Size Blooming Bulbs, 15 cts. each, 2 for 25 cts., $1.50 per Doz.

YELLOW CALLA (Richardia Hastata)—This is identical in all respects to the well-known White Calla, excepting that the flowers are light yellow; very choice and rare. Price 60 cts. each. Good Second Size Bulbs, 35 cts. each.

RED CALLA. (Arum Cornutum)—Red flowers, spotted, with black stems; curiously mottled green and white foliage, palm-like and very handsome. 15 cts. each.

SPECIAL OFFER — 1 Little Gem, 1 Black, 1 Large White, 1 Red, the 4 Callas for only 50 cts., postpaid. Include the Yellow Calla, the 5 for 85 cts.

Black Calla (Arum Sanctum).

THE BOSTON FERN, NEPHROLEPIS EXALTATA.

THIS is one of the finest and best Ferns for growing in pots, vases and baskets; will do well planted in a shady border during summer, and when lifted and taken indoors, makes a splendid window and house plant during Winter. The beautiful fronds grow two to three feet long, and arch over in the most graceful manner. Very popular wherever known. **Price, good strong plants, 15 cts. each, $1.50 per doz., postpaid; larger size, 20 and 25 cts. Extra size, from 5- and 6-in. pots, 50 cts. each, by express, at purchaser's expense.**

The Boston Fern.

Special WINDOW GARDEN Collection

1 Beautiful Boston Fern; 1 New Emerald Feather Asparagus; 1 Weeping Washington Palm, Page 34; 1 New Browallia, Violet blue, Page 2 ; 1 New Flowering Begonia Bijou, Page 30 ; and 1 New Abutilon Savitzii, Page 18.

Set of 6 postpaid, only 65 cts.

NEW EMERALD FEATHER ASPARAGUS.
(ASPARAGUS SPRENGERII.)
See Colored Plate on First Page.

THIS is undoubtedly one of the handsomest and most valuable evergreen trailing plants for the house and conservatory ever introduced. It is especially valuable for pots, vases, baskets, etc., covering all with its beautiful sprays of lovely green feathery foliage, which can be cut freely and are very useful for bouquets, wreaths, and all kinds of floral decoration. It makes a charming ornamental plant for the window or conservatory in Winter, and is equally valuable for vases, baskets and porch boxes in Summer. It is a strong vigorous plant, very easily grown, requires but little care, and keeps on growing, fresh and green, year after year. **Price, 15, 20 and 25c. each ; $1.50, $2.00 and $2.50 per doz., postpaid. Extra size, from 5- and 6-in. pots, 40 and 50c. each, $4.00 and $5.00 per doz., by express.**

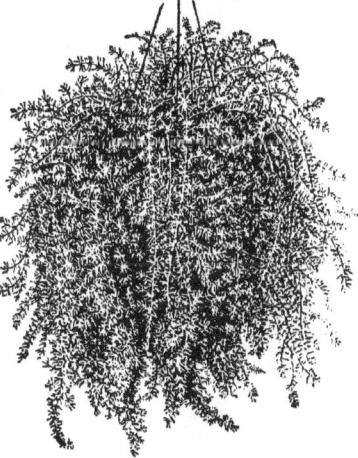

Asparagus Plumosa. Climbing Lace Fern

THIS is a finer and more delicate plant than the Sprengerri, but hardy and easily grown, and very satisfactory for window and house culture. It is an extremely graceful window climber, with bright, green, feathery foliage, as fine as the finest silk or lace. The fronds retain their freshness for weeks when cut, and are greatly admired for floral decoration. An exceedingly beautiful plant for house and conservatory culture, and will thrive nicely in the temperature of an ordinary living-room. Entirely unequalled for the grace and beauty of its lovely, spray-like fronds. **Price, nice thrifty plants, 15 and 20c. each, $1.50 and $2.00 per doz. postpaid. Larger size, 35c. each ; Per doz., $3.50, exp.**

Emerald Feather Asparagus.

CLERODENDRON BALFOURI, A BEAUTIFUL WINTER-BLOOMING PLANT.

This is a slender winter-blooming climbing vine, highly valued for window and house culture, will grow two to three feet high, and can be trailed in any form desired. It is a tremendous bloomer, bearing hundreds of pretty little pear-shaped flowers for months at a time. The flowers are rich bright crimson inside, and creamy white outside, which makes a very striking and attractive combination, always greatly admired. **Good plants, 15 cts. each ; larger size, 25 cts., postpaid.**

Washington Weeping Palm

This is one of the very best Palms for house culture ; hardy, and will thrive in any ordinary living-room regardless of heat or cold, dust or draught ; has elegant fan-shaped leaves, rich dark green, elegantly fringed with white threads ; grows easily, needs no petting or coaxing, always gives satisfaction. Price, good plants, 12 inches and over, 4 to 5 fronds, 15, 20 and 25 cents each, postpaid. Extra size, 18 inches high, with character leaves, 50 cents each, express.

NOTE.--All our Palms are good, sturdy plants, suitable for window and table decoration. They are handsome now, and will grow more beautiful for years.

COCOS WEDDELLIANA.

This is one of the most elegant and graceful of all the smaller palms ; its slender, erect stem is freely furnished with its graceful arching fronds and fine deep green leaves. The Cocos is quite hardy and easily grown, and is highly valued as a very choice table and parlor ornament for vases, fern dishes, etc. In fact, there is nothing else so handsome, and as they are of slow growth, they retain their beauty for a long time, and become permanent household ornaments. Nice plants, 6 to 8 in. high, 3 fronds, 30 cts., postpaid. 10 to 12 in., 5 fronds, 50 cts., by express.

Washington Weeping Palm.

Cocos Weddelliana.

THE Ostrich Feather Palm (Areca Lutescens).

One of the grandest and most beautiful palms for house culture now known. The foliage is rich glossy green with bright yellow stems, full of grace and beauty ; hardy and easily grown ; and grows more beautiful as it grows older and larger. The palms require no special treatment. Will all thrive in parlor or living room. Good, strong plants, 10 to 12 in. high, 3 fronds, 30c. each, postpaid. 15 to 18 in., 4 fronds, 50c. each, by express.

Ostrich Feather Palm.

Umbrella Plant (Cyperus Alternifolia).

The Umbrella Plant somewhat resembles a palm in general style and habit of growth ; grows easily and makes a nice window ornament. Good strong plants, price, 15c. each, postpaid ; larger size, 20c. each.

Latania Borbonica (The Fan Palm).

This beautiful palm is recognized as being one of the handsomest of all, and indispensable to every collection. Always admired ; fine plants 10 to 12 in. high. 4 fronds, 30c. postpaid ; 15 to 18 in., 5 fronds, 50c. each by express.

Umbrella Plant, Cyperus Alternifolia.

Latania.

SPECIAL OFFER. THE FOUR PALMS, AND CYPERUS ALTERNIFOLIA, FIVE IN ALL, POSTPAID, FOR $1.10. SAME, IN LARGER SIZE, PACKED TO EXPRESS HERE, $2.20.

THE KENTIA PALM, Belmoreana.

THE KENTIA PALM, besides being one of the most graceful and ornamental of all palms for the house or conservatory, is also one of the hardiest and easiest to grow. It is of slow growth, but is not affected by the dust and dry air of the house, and will grow and thrive where few other plants would live, and will continue to increase in size and beauty for many years. **Fine thrifty plants, 10 to 12 inches high, 35c. each, postpaid.**

The Rubber Tree. Ficus Elastica.

The India Rubber Tree is well known as one of the very finest plants for table and parlor decoration. Its large, thick, olive green leaves and graceful polished stems make it one of the very finest ornamental plants for the house and conservatory. It stands dust and heat with impunity, and always looks handsome and attractive. The plants we offer are very cheap at the price, and must always be sent by express, as they are too bulky to go by mail. **Fine plants, 15 to 18 inches high, 50 and 75c. each, larger sizes, $1.00 and $1.25 each, express only.**

Kentia Palm.

The New Lady Fern.

Almost every one who sees this beautiful New Fern remarks at once "**It is the most lovely Fern I ever saw**".

And it certainly has few equals among ornamental plants for the parlor and conservatory. It grows low and bushy, and is exceedingly graceful in form, as is shown in our photograph herewith. The fronds have an exquisite moss-like velvety finish, impossible to describe, but of the most delicate beauty imaginable. It is easy to grow, and will thrive and retain its exquisite beauty for years in the parlor or living-room with only ordinary care. It is indeed a beauty. **Nice pot Plants, 25c., postpaid.**

Croton, Aurea Maculatum.

The Crotons are greatly admired for the rich and beautiful coloring of their leaves; some kinds are too tender for ordinary use, but this one will stand all manner of hard usage, and is a most desirable plant for parlor or living room. It grows erect with long narrow deep green leaves, spotted all over with flakes of yellow, looking as if sprinkled with gold dust. Always pretty, and grows prettier every year. **25c. each, 3 for 60c.**

The New Lady Fern.

Genista, or Shower of Gold.

The Genista or Shower of Gold is a charming hard-wooded window plant, of neat bushy form, and almost unequalled for Winter and Spring bloom. Very highly valued for Easter decoration, etc. The flowers are bright golden yellow, delightfully fragrant, and borne in large drooping racemes in such immense profusion, the whole plant seems covered with a shower of golden bloom. It is very easy to grow and absolutely sure to bloom, and is a great favorite wherever known. Set out in flower bed when done blooming, and it will be ready to take in again the next Winter. **15c. each, postpaid. Larger size, 35c., express.**

The Beautiful Wax Plant, Hoya Carnosa

This is not new, but a very beautiful and satisfactory house plant. It is a low climber with brown stems and thick glossy olive green leaves; bears beautiful clusters of exquisite wax-like flowers, creamy white delicately tinted with pink; they are delightfully fragrant and continue beautiful for months at a time. It grows slowly and becomes more beautiful with age. Will do best in a small pot, and should be kept rather dry; easily managed, thrives nicely in the living-room. A popular favorite wherever known. **20c. each.**

Genista.

The Conard & Jones Co., West Grove, Pa.

ROSE AND FLOWER GROWERS.

We ask especial attention to our

NEW

Rambler
Roses

FOUR VARIETIES.
FOUR DIFFERENT COLORS.
FOR FALL PLANTING.

The Crimson Rambler, intense dazzling crimson.
The White Rambler, pure pearl white,
The Pink Rambler, clear, bright pink.
The Yellow Rambler, fine golden yellow.
All perfectly hardy, fine for Porches, Verandas, Trellises, etc. Immensely popular.

DESCRIPTION—The Rambler Roses are strong, vigorous climbers, growing ten to twelve feet high in a season. They bear immense clusters of beautiful fragrant flowers, and will soon cover the whole side of a house with a sheet of lovely bloom. They are the grandest climbing roses yet introduced, perfectly hardy, need no protection and grow more beautiful every year. If you have not got them, do not fail to order them now.

Strong Mailing Plants, from 3-inch pots. —15 cts. each, 2 for 25 cts.; 4 for 50 cts.; $1.50 per doz., postpaid.

Two-year Ramblers, from open ground.—When sent by mail, postpaid, 35 cts each ; set of 4 for $1.25 By express, 25 cts each, 4 for $1.00 ; $2.50 per doz.

Extra Size, two-year Ramblers.—Field-grown dormant bushes, 65 cts. each, 2 for $1.25 ; set of 4 packed to express here, for $2.00 ; $5.00 per doz.

Three-year Ramblers.—Extra large, dormant field-grown bushes, 75 cts. each, 2 for $1.40 ; 4 for $2.75, by express, purchaser paying charges.

New Double White Hardy Climbing Rose, THE

Royal Cluster,

Bears immense clusters of pure white medium size roses, quite double and fragrant, and recommended as the best all-round Pure White Hardy Climbing Rose to date. Entirely hardy. **Strong mailing plants, 15c. each, 2 for 25c.; $1.50 per doz., postpaid. Two-year size, 30c. each, 3.00 per doz., by express,**

COPYRIGHTED BY
CONARD & JONES CO.
1898

Two Ramblers, One Crimson and One White.

Nine Choice New Geraniums
Strong Plants from 3-in. Pots

NEW DOUBLE GERANIUM, ROSEMAWR—A sport from Mrs. Taylor. Finely crinkled deep green leaves, with rich chocolate band, and large compact heads of bright rose-pink flowers, with creamy white centre. A bushy, compact grower, and constant bloomer. 15 cts. each.

NEW GERANIUM, PEACH BLOSSOM—An elegant new variety, bearing large heads of beautiful flowers, rich peach-blossom red, very distinct and handsome, a good thrifty grower, free bloomer. 15 cts.

NEW HOUSE GERANIUM, MARS—This is a beautiful and handsome geranium, especially recommended for house culture; large single flowers, lovely peach pink, with rich orange-red centre; an immense bloomer, covered with flowers nearly all the time, foliage prettily marked with dark brown, very striking and attractive. 15c.

NEW GERANIUM, MAD. BRUANT—A splendid new variety, entirely distinct from all others. White, elegantly veined with rich carmine; florets edged and striped with crimson lake, both flowers and trusses are very large and beautifully formed; plant is a healthy grower and free bloomer. Mad. Bruant is one of the most remarkable new geraniums ever sent out, magnificent. 15 cts.

MLLE. MARIE HERBERT, or APPLE BLOSSOM—Large single flowers, sometimes nearly two inches across, and borne in grand trusses; color white, elegantly shaded and marked with rose, strongly reminding one of a cluster of apple blossoms; handsome foliage and free bloomer. 15 cts. each.

NEW GERANIUM, COLOSSUS—Grand semi-double flowers, two and a half inches in diameter, almost like roses, immense trusses; color, a rich shade of rosy crimson; very free bloomer, one of the very largest and finest double geraniums ever seen. 15 cts.

New Geranium Rosemawr

NEW DOUBLE GERANIUM, ALPHONSE RICAUD—Bright, orange red, in trusses of largest size; wonderfully beautiful and almost unequalled. 15 cts. each.

ALPINE BEAUTY—Handsome compact grower, extra large, double; pure white flowers, constant and profuse bloomer, one of the best. 15 cts.

NEW PANSY PELARGONIUM, or "LADY WASHINGTON GERANIUM"—We are glad to offer this elegant variety, for scarcely any plant grown in the window garden will give better returns with so little care. Placed in a light sunny window, it will be literally covered with flowers for weeks or months, continuing in bloom nearly all summer; flowers pink and crimson, with dark blotches, sometimes feathered white, very handsome and showy. 15 cts.

SPECIAL OFFER: any 3 for 35c. Set of 9, postpaid, $1.00

Variegated Leaved Geraniums
These are highly valued for Window Culture

MADAME SALLEROI Makes a thrifty, compact, bushy plant, only six to twelve inches high; leaves, light silvery green, with wide white edge. 15 cts. each.

MRS. POLLOCK—A beautiful tri-colored variety; leaves, rich metallic bronze, banded with scarlet, and edged with golden-yellow, bright scarlet flowers. 15 cts.

MOUNTAIN OF SNOW—Centre of leaves bright, glossy green, widely bordered with pure white, pretty pink flowers, fine for pots. 15 cts. each.

Set of 3, postpaid, for only - - - 35c.

New Chrysanthemums
Strong Plants from 3-inch Pots
Some in Bud and Bloom - -

BUFF HAIRY, OSTRICH PLUME—A grand Japanese variety, beautiful buff-yellow, immense globular flowers, covered all over with fine feathery plumes. 12 cts. each.

CHILD OF TWO WORLDS, OSTRICH PLUME—Pure snow-white, elegantly plumed, very fine. 12 cts. each.

CHARLOTTE—Beautiful, pure pearl-white flowers; immense size and finely formed. 12 cts. each.

DRAGON FIRE—Extra large full flowers, deep glowing red, with golden yellow tips. 12 cts. each.

GLORY OF THE PACIFIC—A beautiful clear pink fine variety of magnificent size and depth. 12 cts. each.

LOUIS BOEHMER, OSTRICH PLUME—Exquisite silvery pink, very fine. 12 cts. each.

MADAM F. CAYEUX. OSTRICH PLUME—Bright, rich, ruby-red, petals tipped with old gold, and elegantly plumed. 12 cts.

MRS. A. J. DREXEL—Large flowering; color, crimson lake. 12c.

MRS. PERRINE—Extra large, incurved, globular flowers; bright, clear rose-pink, a prize winner. 12 cts. each.

MISS M. M. JOHNSON—Deep golden-yellow, beautiful, early and abundant bloomer. 12 cts. each.

MODESTA—Rich velvety yellow, unsurpassed in size; incurved and slightly whirled form, with full, high centre. 12 cts.

NIVEUS—Grand snow-white globular flowers, good grower and bloomer. 12 cts. each.

PINK IVORY—Grand clear pink, constant bloomer. 12 cts.

Special Offer 3 for 30c., set of 13 postpaid, for only - - - - $1.10

Winter-Blooming Chrysanthemums

NEW ORANGE GUAVA.

New Orange Guava.

Yellow Catley Guava.

THIS is a delightful tropical fruit which here makes an elegant house plant, rivalling the Otaheite Orange in its beauty and a nice companion for it. Makes a neat, handsome plant, with thick glossy-green leaves, and pure white fragrant flowers. The fruit is nearly the size of a walnut and rich golden yellow. The flavor is sweet and delicious, begins to bloom when quite small, and bears both flowers and fruit at the same time ; is easily grown and a most charming and interesting plant for house culture. **The very best Guava Jelly is made from this variety. Fine strong plants, 20c. each, 3 for 50c, postpaid.**

FLORIDA LIME—A beautiful house plant; hardy and easily grown; very similar in appearance to the orange; thick, glossy green leaves, small waxy white flowers, borne in clusters and deliciously sweet, small, round yellow fruit ; very fragrant. A nice companion for the orange. 15c. each, the two for 30c

Two-year Hardy Field-grown Roses. FOR FALL PLANTING IN OPEN GROUND.

EMPRESS OF CHINA—Hardy ever-blooming climber, deep rose pink flowers. Strong two-year plants, 25c. each, $2.25 per doz., by express.

HARRISON'S YELLOW—The best hardy yellow rose; clear golden yellow, entirely hardy, fine for planting with other ornamental shrubs. Strong field-grown plants, 25c. each; $2.25 per doz., by exp.

RUBY QUEEN—A good companion for May Queen, equally hardy and desirable ; clear bright red with white centre. Strong two-year bushes, 25c. each, $2 25 per doz., by express.

MAY QUEEN—New hardy climbing rose ; clear coral pink, double and very fragrant. Strong field-grown plants, 25c. each, $2.25 per doz., by express.

THE GARLAND—A fine English Running Rose, valued for training over walls, fences, embankments, etc ; blooms in clusters, pure white, very full and fragrant. Two year bushes, 25c. each, $2.25 per doz ; by express.

PURE WHITE MEMORIAL ROSE. (Wichuraiana)—For cemetery and park planting. Two year bushes, 25c. each; $2.25 doz., express.

When sent by Mail, 30 cts. each, 2 for 50 cts., $3.00 per dozen, postpaid.

Hardy Flowering Shrubs. STRONG FIELD=GROWN PLANTS FOR FALL PLANTING.

Strong Mailing Plants, except where noted, 15 cts., each ; 5 for 50 cts.; 10 for $1.00 Postpaid. Two-year Size, by express, 20 cts. each ; 6 for $1.00, Per Doz., $1.75.

WHITE FRINGE CHIONANTHUS—Lovely pure white fringe-like flowers, very beautiful. 15c., postpaid

WEIGELIA—Variegated leaved. Leaves bordered creamy white; flowers, blush pink. 15c. each, postpaid.

ALTHEA—Variegated leaved; leaves bordered with creamy white; hardy and desirable. 15c. each.

DEUTZIA GRACILIS—Low bushy grower, pure white, very hardy. 15c. each, postpaid.

DEUTZIA, PRIDE OF ROCHESTER—Double flowers; white, shaded pink. 15c., postpaid.

FORSYTHIA VIRIDISSIMA (Golden Bells.)—Bright yellow flowers very early in Spring. 15c., postpaid.

HYDRANGEA GRANDIFLORA—Grandest of all hardy shrubs. 10, 15 and 20c. each, according to size, postpaid, all will bloom the first season. Three-year size 40c., each, $4.00 per doz., express.

SPIREA NEW CRIMSON. (Anthony Waterer.) —Blooms profusely in Spring, and at intervals all through the season; rich rosy red, hardy. Best hardy shrub recently introduced. See full description in Spring Guide. 15c., postpaid. Three Year size, 30c.; $3.00 per doz., by express.

SPIREA, NEW BLUE—(Caryopteris)—Lovely sky blue flowers from August on; best blue flowering garden shrub we know. 15c., postpaid.

SPIREA PRUNIFOLIA.—Bridal Wreath. Immense bloomer; pure white double flowers, like little roses, perfectly hardy. 15c., postpaid.

ALTHEA—New double white, Jean d'Arc; large double flowers, pure snow white; very hardy. 20c., postpaid. Two-year size, 25c.; $2.50 per doz., by express.

Hardy Climbing Vines.

AMPELOPSIS VEITCHEII—Japan or Boston Ivy. The best and most beautiful hardy climbing vine for covering the walls of houses, churches, schools mills, etc. Good strong plants. 15c. each; $1.50 per doz., postpaid. Two-year size, 20c.; $2.00 per doz., by express, at purchaser's expense.

CLEMATIS. Japan Sweet-Scented—(Paniculata.) Pure white, hardy and very fine. 15c.; $1.50 per doz., postpaid.

WISTERIA Chinese Blue—15c. each, postpaid. Two-year size, 25c ; $2.00 per doz., by express.

SWEET-SCENTED HONEYSUCKLES—Chinese Sweet Scented, Evergreen Sweet-Scented, Halliana Sweet-Scented, Red Coral, and Golden Leaved Honeysuckle. 15c. each, two for 25c., set of five for 50c.

FIELD-GROWN CARNATIONS

BLOOMING SIZE.

Copyrighted by
CONARD & JONES CO
1898

CARNATION PINKS

Are the sweetest and most beautiful flowers you can possibly have for winter bloom, and it is easy to have them in abundance if you get **Our Strong Field-grown Clumps**, which are specially prepared for Winter-flowering, and will break into full bloom almost as soon as potted, and continue bearing their lovely flowers as long as kept in growing condition. We can send small orders by mail at prices given, but it is better, whenever convenient, to have these large plants sent by express, as they carry in better condition if some earth is left on the roots, and they are not packed too close in the boxes. The following are the very best varieties for Winter flowering.

STRONG FIELD-GROWN CARNATIONS FOR IMMEDIATE BLOOM.

ROSE QUEEN—Fine rose-pink flowers. Very full and regular, delightfully scented, one of the best. 20 cts.

PURITAN—Large full flowers, pure white and very sweet, early and profuse bloomer. 20 cts.

PORTIA—Rich glowing crimson, large bold flowers, very sweet, one of the best. 20 cts.

KITTY CLOVER—Exquisite yellow flowers, very free bloomer, deliciously scented, very beautiful. 20 cts.

HELEN KELLER—A splendid fancy variety; pure white, beautifully variegated with rich scarlet, very handsome. 20 cts.

THOMAS CARTLEDGE—Bright coral pink, fine full flowers, borne on long, stiff stems; very sweet, one of the best. 20 cts.

WILLIAM SCOTT—Large elegantly fringed flowers, bright clear pink, very beautiful and sweet, a good grower and constant bloomer. 20 cts.

SPECIAL OFFER: 3 for 50c.; 7 for $1.00, postpaid. Express, at purchaser's expense, 15c. each, $1.25 per doz.; $8.00 per 100.

New Garden Fruits FOR Fall Planting

New Dwarf Bismarck Apple. We again take pleasure in recommending this wonderful New Dwarf Apple for planting in gardens and yards where fine fruits are desired, but not room for large trees. Has now been fully tested in this country, and proved to be a most valuable variety; very remarkable for its dwarf growth, and surpassing all others in early bearing. Little trees, not over a foot or two high, will frequently bear large apples a few months after planting, and it seems certain they can be expected to bear abundantly in a very short time. They are so dwarf they can easily be grown in pots or boxes if desired, but are especially recommended for small gardens and yards where there is not room for large trees. You can grow them in any vacant corner, and have some delicious fruits of your own. The apples are of large size, beautiful rich golden-yellow, streaked with red; they ripen early, and are of most delicious quality. We can recommend the Bismarck Apple as a most valuable variety for situations noted above. May be planted only three or four feet apart along fences or buildings, and should soon bear abundantly.

Price, nice One-year Trees, 25c. each, 6 for $1.25, $2.50 per doz., postpaid. Larger size, 35c. each, $3.50 per doz. by express at purchaser's expense.

3 Luscious Japan Plums

Too much cannot be said in favor of these delicious Japan Plums. They make nice little low-headed trees, and will usually mature a crop the second year after planting. Scarcely any other fruit trees will bear so quickly and abundantly, and few fruits are more delicious in flavor and quality. They are especially desirable for planting in gardens and yards where there is not room for large trees. They are easy to grow, require no special treatment, and may be depended upon to bear regularly every year. We offer three of the very best varieties.

NEW JAPAN PLUM, THE HALE—Noted for large size, great productiveness, and excellent quality. Orange yellow, mottled cherry red, exceedingly beautiful and attractive. Nice one-year trees, mailing size, 15 cts. each, $1.50 per doz., postpaid. Two-year trees, 25 cts. each, $2.50 per doz. by express.

ABUNDANCE—Beautiful lemon yellow, with bright cherry red bloom; very large egg-shaped, delicious flavor, most abundant and regular bearer. Nice one-year trees, 15 cts. each, $1.50 per doz., postpaid. Two-year trees, 25 cts. each, $2.50 per doz. by express.

RED JUNE—Highly valued for early ripening and excellent quality, deep purplish red, very handsome and attractive, and the best of the early sorts. Nice one-year trees, 15 cts. each, $1.50 per doz., postpaid. Two-year trees. 25 cts. each, $2.50 per doz. by express.

The Three Varieties, one each, mailing size, only 40c., postpaid. The three in two-year size, 65c. by express.

ENGRAVED FROM A PHOTOGRAPH.

New Dwarf Bismarck Apple.

New Strawberry-Raspberry Or Tree Strawberry

This is one of the most beautiful fruits ever seen; berries the size and shape of the largest strawberries; bright, rich, shining scarlet, with an exquisite bloom; makes lovely jam, jellies and tarts, having a delicious flavor entirely different from any other fruit; bears the first season, and gets larger and stronger every year. Came from Japan, and tested for two years here; the bush grows from eighteen inches to two feet high, is entirely hardy and will do well everywhere, regardless of heat and drought. We cordially recommend it. 10 cts. each, 3 for 25 cts., 75 cts. per doz.

New Gooseberries For... Fall Planting

AMERICAN TRIUMPH—The best American Gooseberry. An enormous bearer, and remarkably free from rust and mildew; large size, with their very thin skin and fine flavor, always scarce and high. 20 cts. each, $2.00 per doz., postpaid; $1.65 by express. Two-year plants, $2.50 per doz. by express.

RED JACKET GOOSEBERRY—Fruit, large dark rich red, excellent quality, and a tremendous bearer; hardy and highly recommended. 20 cts. each, $2.00 per doz., postpaid; $1.65 by express. Two-year size, $2.50 per doz. by express.

New Japan Plum, Abundance,

Japan Giant Chestnut

Of the many good fruits introduced from Japan, none are more worthy than the Japan Giant Chestnut. The tree is quite ornamental, hardy and very productive, of dwarf compact habit, and begins to bear so young that the little trees are frequently loaded with fruit while in the nursery rows. The nuts are of enormous size, about four times as large as the common chestnut, and always sell at a high price. The Japan Giant Chestnut is a tree that everyone should plant. It is the best chestnut yet introduced, and will give immense satisfaction. Price, postpaid, · · · · · · · **25c. each.**

Some Choice Flower Seeds for Autumn Sowing

Pansy seeds are planted largely in the open ground in the Fall for Spring bloom, and we offer the best kinds for this purpose. The other seeds we offer are for planting in pots and boxes indoors, for Winter and Spring bloom. Many handsome window gardens are all grown from seeds.

OUR GOLD MEDAL PERFECTION PANSIES are not excelled in size and beauty by any offered in this country. They are the best. Packet, 50 seeds, 6 cts ; three packets for 15 cts.

ROYAL PRIZE—A famous mixture of all the best varieties of English, French and German Pansies. Pkt. 100 seeds, 6 cts.; 3 pkts. for 15 cts.

NEW GIANT TRIMARDEAU—Giant Pansies, richly spotted with bright and striking colors. Pkt. 100 seeds, 5 cts.; 3 pkts. for 12 cts.

IMPERIAL GERMAN SPLENDID MIXED—More than 50 different shades and colorings of the finest German pansies. 100 seeds, 6c.

NEW PANSY METEOR—Bright canary yellow, with brown and purple markings, and gold or silver edging. 100 seeds, 5 cts.

NEW PEACOCK PANSY—Lovely Peacock colors. 100 seeds, 5 cts.

NEW PANSY, VICTORIA RED—Deep rich red, with narrow gold border. 100 seeds, 5 cts.

CONARD'S SUNSHINE PANSIES—Bright rich colors, very handsome. 100 seeds, 5 cts.; 3 pkts. 12 cts.

SNOW QUEEN PANSY—Pure satiny white, with yellow dot in centre, very popular. 100 seeds, 5 cts.

GOOD QUALITY MIXED PANSIES FOR BEDDING—100 seeds 3 cts.; ⅛ oz., 15 cts.; ¼ oz., 25 cts.

Complete Set of 10 pkts. of Pansies, 35 cts.

NEW JAPANESE MORNING GLORY—These are fine for winter bloom. They flower quickly, and if given string or wire to climb on, will soon cover your window with a mass of vines and flowers. Pkt. 6 cts., oz. 20 cts.

BEGONIA, TUBEROUS-ROOTED DOUBLE—Very fine. Pkt. 12 cts.

BEGONIA, TUBEROUS-ROOTED SINGLE—Choice mixed. Pkt. 10 cts.

BEGONIA VERNON—A grand new perpetual-blooming sort. Pkt. 6 cts.

BEGONIA HYBRID—Finest colors mixed, grand for window culture. Pkt. 6 cts.

CALCEOLARIA DWARF HYBRIDS Large flowering, mixed colors. Pkt. 10 cts.

CARNATION, MARGARET—Fine early-blooming carnations. Pkt. 6 cts.

FERNS—Mixed, many sorts. Pkt. 6 cts.

CARNATION, PERPETUAL OR TREE—Best greenhouse varieties. Pkt. 10 cts.

CINERARIA HYBRID—Large flowering, finest colors mixed. Pkt. 10 cts.

CYCLAMEN PERSICUM—Choicest colors mixed. Pkt. 8c.

CYCLAMEN GIGANTEMUM—Pkt. 10 cts.

COLEUS—Choice mixed hybrids. Pkt. 6 cts.

GERANIUM ZONALE—Choice mixed from the finest new varieties. Pkt. 10 cts.

GERANIUM, APPLE-SCENTED—Very delightfully scented leaves. Pkt. 8 cts.

GLOXINIA, CHOICEST MIXED HYBRIDS—Magnificent. Pkt. 10 cts.

PRIMULA SINENSIS FRIMBRIATA—(Fringed Chinese Primrose,) large flowering, fine, splendid mixed. Pkt. 8c.

PRIMULA OBCONICA—A profuse blooming Primrose, flowers pure white, shading to lilac, fine for winter bloom. Pkt. 6 cts.

SMILAX—One of the finest climbing vines for window. Pkt. 4 cts.

NASTURTIUM, DWARF OR TOM THUMB—There are few flowers equal to Nasturtiums, for pot culture and winter blooming, they are superb. Pkt. 25 seeds, 6 cts., oz. 15 cts.

NASTURTIUM, TALL SWEET SCENTED—The finest mixed. Pkt. 25 seeds, 6 cts., oz. 15 cts.

FOUR CHOICE NEW WINDOW PLANTS

ACALYPHA MOSAICA—A handsome new foliage plant for pot culture. Thick glossy leaves, looking as if varnished, and beautifully variegated with red and old gold. 10 cts. each.

GIANT SWEET-SCENTED SNAP DRAGON—A neat, handsome plant, throwing up dozens of graceful spikes bearing lovely creamy white flowers; very sweet and blooms all the time. 10 cts. each.

RUELLIA MAKOYANA—A rare new house plant, highly recommended, rich velvety green foliage, beautifully veined with white; elegant tubular flowers, carmine, rose, abundant bloomer. 10 cts. each.

PELLONIA MEXICANA—A pretty low-growing plant for pots and vases, ornamental; foliage, olive green, marbled with silvery gray. 10 cts. each.

Set of Four, only   .   .   **35 cts.**

CHOICE FUCHSIAS FOR WINTER BLOOM

NEW FUCHSIA, MADAM BRUANT—One of the most magnificent Fuchsias ever introduced ; immense flowers, perfectly full and double, lovely weeping tree habit, rich purple and crimson. 10 cts. each.

NEW DOUBLE WHITE FUCHSIA, GLORIE DE MARCHE—A magnificent variety, makes a neat compact plant, branches freely, and is loaded nearly all the time with exquisite double flowers of largest size, and of pure snowy whiteness, extra fine every way. 15 cts. each.

JUPITER—Extra large and beautiful; long deep tubes and wide handsome corolla, rich bright crimson. 10 cts

TRAILING QUEEN—A lovely trailing variety, fine for vases, etc.; wine color, with carmine sepals. 10 cts.

WAVE OF LIFE—Beautiful golden foliage, crimson tube and sepals, and fine purple corolla. 10 cts. each.

SPECIOSA—The true winter-blooming fuchsia. Blooms all Fall and Winter, one of our prettiest winter-blooming plants. 10 cts. each.

Complete Set of Six, postpaid, for   .   .   **50 cts.**

NEW WEEPING LAN

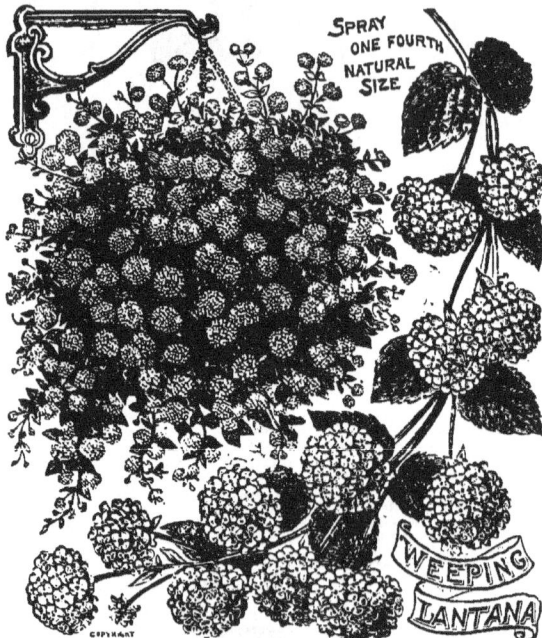

SPRAY ONE FOURTH NATURAL SIZE

WEEPING LANTANA

New Weeping Lantana.

Mrs

THIS pre
neat,
larly
bloom in p
flowers are
in clusters
will bloom
in growing
comes, set
it will bl
bed of it
of bloom
a great de
2 for 2
$1.25 pe

New /
"Prin

A very
dwa
exce
height a
its large
ers, whic
deep blue
bloomer,
the time
4 for 5(
postpaid

SPECIAL WINDOW GARDEN SET.

1 New Weeping Lantana; 1 Ageratum Princess Pauline; 1 Royal Purple Bougainvillea; 1 Umbrella Plant; 1 Grevillea Robusta, Fern Leaf Tree; 1 Acalypha Mosaica, Autumn Leaf Plant; 1 Winter Blooming Fuchsia, Speciosa. Set of seven choice house plants, only **50 cts.**

Grevillea Robusta.

Silk Oak, or **Fern Leaf Tree.** An elegant decorative plant for the house or conservatory ; rapid, growth, finely cut foliage, rivalling a rare fern or palm. Strong, thrifty plants 15 cts. each,

Swainsonia R

plant exactly like Swainsonia Alb
is bright, rosy pink. Very ha
bloomer. **15 cts. each, 2 for 2**

SCILLA

ONE of the hardiest and most
equally valuable for Wint
for bedding in open groun
always sure to bear an abunda
flowers. Indoors it will bloom f
and is one of the very earliest
ded out ; very beautiful. **3 f(**
doz., 75 cts. per 100, postpa

Scilla Siberica.

www.ingramcontent.com/pod-product-compliance
Lightning Source LLC
Chambersburg PA
CBHW022030190326
41519CB00010B/1649